T0325124

Fire Performance of Thin-Walled Steel Structures

Fire Performance of Thin-Walled Steel Structures

Yong Wang

Mahen Mahendran

Ashkan Shahbazian

CRC Press
Taylor & Francis Group
Boca Raton London New York

CRC Press is an imprint of the
Taylor & Francis Group, an **informa** business

CRC Press
Taylor & Francis Group
6000 Broken Sound Parkway NW, Suite 300
Boca Raton, FL 33487-2742

© 2020 by Taylor & Francis Group, LLC
CRC Press is an imprint of Taylor & Francis Group, an Informa business

No claim to original U.S. Government works

Printed on acid-free paper

International Standard Book Number-13: 978-1-138-54085-9 (Hardback)

**Visit the Taylor & Francis Web site at
http://www.taylorandfrancis.com**

**and the CRC Press Web site at
http://www.crcpress.com**

Contents

Preface

The behaviour of thin-walled structures is complex and fire resistance of structures is a niche subject. Therefore, fire resistance of thin-walled structures has attracted a relatively low level of interest and investment from the research community. As a result, the current design guidance for the evaluation of fire resistance of thin-walled structures is rudimentary, still dominated by fire resistance testing.

However, thin-walled steel structures are increasingly used in building construction and large numbers of thin-walled steel structure products are developed owing to flexibility of the manufacturing methods. Research and development of thin-walled structures by fire testing will no longer be tenable due to prohibitive cost and limited scope of application, and methods of calculation will become essential.

Developing calculation-based methods for fire resistance of thin-walled structures demands a thorough understanding of the fundamentals of a number of subjects and availability of reliable input data for material properties. To aid this process of development, this book provides a single source of information which at present is scattered in academic articles. This book attempts to provide an authoritative account of the latest developments in the different aspects of the subject matter, including fire resistance requirements, behaviour of thin-walled structures in fire and measures of improving fire resistance of thin-walled structures, fire behaviour, simplified heat transfer modelling to obtain temperatures in thin-walled structures, temperature-dependent thermal and mechanical properties of materials, and latest developments in evaluating the load carrying capacity of thin-walled steel structures at elevated temperatures by calculation.

The book is jointly written by two research groups who have had a long and distinguished record of extensive research on fire resistance of thin-walled steel structures and are active in developing calculation methods. The authors hope that this book will be an indispensable reference to researchers of this increasingly important field. We also believe that the book will be valuable to fire protection engineers who want to optimise fire resistant design of thin-walled steel structures through developing a firm understanding of the first principles of the subject matter, and specialist manufacturers of thin-walled steel structures in their efforts to develop more efficient products by helping them understand the factors that control fire resistance of thin-walled steel structural systems.

Transforming our motivation to action of writing this book was greatly helped by the decision of our publisher, Taylor & Francis, to launch the CRC Focus book format which is ideal for presenting a complete treatment of this specialist subject in a concise manner. Writing a book in our spare time for a deadline long in the future was never going to make it a priority until reaching the last minute, which would surely be a disaster. We were spared of this crisis by Gabriella Williams, who regularly checked that we were making progress according to our plan.

Yong Wang,
Mahen Mahendran,
Ashkan Shahbazian

Authors

Yong Wang is professor of structural and fire engineering at the University of Manchester, UK, where he leads the Structural Resilience research group. He is author of *Performance-Based Fire Engineering of Structures* and *Steel and Composite Structures: Behaviour and Design for Fire Safety*, also published by Taylor & Francis.

Mahen Mahendran is professor of structural engineering at Queensland University of Technology, Australia, where he leads the Wind and Fire research lab. He has served for more than 35 years at six universities.

Ashkan Shahbazian is an adjunct assistant professor at the Institute for Sustainability and Innovation in Structural Engineering, University of Coimbra, Portugal and Head of Research and Development at the Iranian Society of Structural Engineering.

Authors

List of notations

LOWER CASE

b	width
d	depth of steel cross-section
e	eccentricity/moisture content
f	a constant
f_y	yield strength of steel
h	convective heat transfer coefficient
k	thermal conductivity
t	time
x,y,z	coordinates

UPPER CASE

A	area
C_p	specific heat
E	Young's modulus
L	length/span/height
N	compression resistance
P_n	design resistance under compression
$\dot{Q}$	heat
R	thermal resistance
T	temperature
T_f	fire/furnace temperature
T_s	surface/steel temperature
W	width of panel

SUBSCRIPT

0	ambient condition
b	buckling
cap	related to capacitance
cond	conduction
conv	convection
cr	critical
d	distortional
e	effective/Euler buckling
Ed	design effect
f	flange/fire
l	local
rad	radiation
Rd	design resistance
ref	reference
s	surface
y	yield

GREEK LETTER

α	coefficient of thermal expansion
ε	emissivity
ρ	density
σ	Stefan-Boltzmann constant
λ	slenderness
Δ	difference
Σ	sum

Introduction
Fire Safety Requirements and Implications for Thin-Walled Steel Construction

1

This chapter presents the general requirements of fire safety and, in particular, fire resistance, the implications for thin-walled steel construction, methods of evaluating fire resistance of thin-walled steel structures and the contents of this book.

1.1 THIN-WALLED STEEL STRUCTURES

Thin-walled steel structures (alternatively referred to as light gauge steel structures, light steel framing, lightweight framing) are increasingly used in building construction worldwide, not only as secondary structural members such as sheeting and purlins, but also as primary load-bearing members forming walls and floors. For example, Figure 1.1 shows an example of thin-walled steel studs as part of the wall construction of a residential building. Thin-walled steel structural components are normally cold-formed using 0.55–6.35 mm thick low- and high-strength steels (nominal yield strength in the range of 250–550 MPa at ambient temperature). C-section profiles are commonly used, however, with significantly

FIGURE 1.1   Thin-walled steel studs in a residential building. (Courtesy of Metek, Elmshorn, Germany.)

advanced manufacturing technologies; other innovative and structurally efficient steel sections are being produced and used to suit specific applications.

Owing to their lightweight, a variety of construction methods can be used to build thin-walled steel walls, floors and roofs. They include:

1. Stick building: Whereby individual studs are framed on site, one at a time. Interior insulations and panels are then installed to form the complete walls and floors. This method gives flexibility on site, but is time consuming.
2. Panel construction: Thin-walled steel members are pre-assembled in a factory, together with interior insulations and panels, and then transported and erected on site, for example, wall and floor panels and trusses. This method reduces the on-site construction time, improves material efficiency and construction quality.
3. Volumetric construction: Whole rooms are constructed in a factory, together with fittings and electrical/mechanical services, and then transported to and connected on site. This method of construction further improves on-site construction speed and material efficiency. Although volumetric construction has been promoted for some time, often in the context of modern methods of construction to help

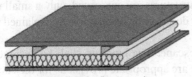

18 mm chipboard
Light steel joists
100 mm mineral wool between
joists
12.5 mm plasterboard

• Example of a floor panel

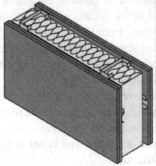

2 layers of 12.5 mm plasterboard
Resilient bars
Light steel studs with mineral wool
between
Resilient bars
2 layers of 12.5 mm plasterboard

• Example of a wall panel

FIGURE 1.2 Examples of thin-walled steel structure construction. (From Steel Construction Institute (SCI), Fire resistance of light steel framing, SCI Publication P424, Steel Construction Institute, Ascot, UK, 2019. With permission.)

solve the problem of lack of qualified skills, its market share is still relatively small. User confidence in the quality and flexibility of this type of construction appears to be an issue.

Whatever the method of construction, the final product is a panel with enclosed thin-walled steel members, with or without interior insulation. Figure 1.2 shows examples of thin-walled steel structural floors and wall panels. It is the fire performance of the complete panel assembly that will be the topic of this book.

1.2 FIRE SAFETY REQUIREMENTS AND THEIR IMPLICATIONS FOR THIN-WALLED STEEL STRUCTURES

Although different countries have different specific requirements for fire safety of buildings, the requirements can generally be grouped into the following five sets, as in the UK's Approved Document B to the Building Regulations (BR2010).

However, as far as thin-walled steel structures are concerned, only a small number of these requirements affect their design and construction, as explained next.

- B1 Means of warning and escape: The building shall be designed and constructed so that there are appropriate provisions for the early warning of fire, and appropriate means of escape in case of fire from the building to a place of safety outside the building capable of being safely and effectively used at all times. Means of escape considerations include local means of escape from the initial fire enclosure when the fire is at the pre-flashover stage and global means of escape after the fire in the initial fire enclosure has developed into the post-flashover stage. For thin-walled steel structures, as for any other types of structures, the post-flashover fire stage governs their fire safety design and construction, which is the subject of fire resistance (B3, see later). As for detailed assessments for means of warning and escape, this requirement does not affect and is not influenced by the use of thin-walled steel structures.
- B2 Internal fire spread (Linings): To inhibit the spread of fire within the building, the internal linings shall:
  - Adequately resist the spread of flame over their surfaces
  - Have, if ignited, a rate of heat release or a rate of fire growth, which is reasonable in the circumstances.

  Here, 'internal linings' mean the materials or products used in lining any partition, wall, ceiling or other internal structure. The lining materials on thin-walled steel structures are generally non-combustible, such as gypsum plasterboards (see Figure 1.2).
- B3 Internal fire spread (Structure): The building shall be designed and constructed so that, in the event of fire, its stability will be maintained for a reasonable period. This is to ensure that the fire does not spread out of the initial compartment where the fire breaks out.

  This requirement is concerned with fire resistance, which is the ability of the construction to prevent fire spread by achieving adequate insulation, integrity and load-bearing performance. Insulation performance is achieved if temperature on the unexposed surface of the construction is low, defined as no more than 140°C on average, and no more than 180°C at the maximum above the ambient temperature. Adequate integrity performance ensures that the fire does not burn through the construction. Load-bearing capacity is necessary to ensure that the structure of the construction does not fail/collapse, thereby leading to fire spread from one compartment to another.

Thin-walled steel structures usually form the walls and floors of fire-resistant construction. Therefore, they need to achieve all of the above three fire resistance requirements. At present, integrity of construction in fire still cannot be quantified by calculations; therefore, detailing instructions according to the instructions of specialist suppliers of the construction, fire testing or expert opinions should be adhered to. This includes penetrations for services and cavity spaces in the construction.

It is possible to demonstrate fulfilment of the insulation and load-bearing requirements by calculations, and these are the main focuses of this book.

- B4 External fire spread: This performance requirement mainly deals with lining materials on the external surface of the construction, similar to requirement B2 for the internal lining materials. Gypsum boards are non-combustible materials. If other types of lining materials are used on top of the plasterboards, they should meet the requirements of the regulations. However, these requirements are the same for thin-walled steel structures as for other construction types. It does not impose any additional requirement on thin-walled steel structures.

- B5 Access and facilities for the fire service: This requirement is about providing reasonable facilities to assist firefighters and making reasonable provisions within the site of the building to enable firefighting appliances to gain access to the building. However, as thin-walled steel structures are usually enclosed by gypsum plasterboards to form walls and floors, one concern is that firefighting water may soak up the gypsum plasterboards. Water itself is beneficial to fire resistance as evaporation of water takes a lot of heat away from the construction, thus slowing down temperature rise. The concern is the amount of firefighting water increasing weight of the plasterboards and reducing mechanical properties of the connection, thereby causing premature collapse of the floor. However, because the amount of firefighting water is low, in comparison with weight of the plasterboards, and there is no historical evidence of firefighting water causing plasterboard failure in fire, this issue is minor.

Therefore, the main implications of fire safety of buildings on thin-walled steel construction are fire resistance requirements. Of the three specific fire resistance requirements of insulation, integrity and load-bearing, meeting the integrity requirement is not by calculations. Since this book is to present calculation methods, it will only cover insulation and load-bearing.

1.3 DETERMINATION OF FIRE RESISTANCE OF THIN-WALLED STEEL STRUCTURES

The fire resistance of thin-walled steel construction can be demonstrated in two generic ways, by standard fire resistance testing or by calculation. Testing is necessary because thin-walled construction involves many systems and details. However, standard fire testing is expensive and the results can only be applied to the particular construction system tested.

The scope of calculation methods can range greatly. Options include:

1. Extension based on fire test results.
2. Limiting temperature method.
3. Extension of ambient temperature design methods.
4. Simplified heat transfer and load-carrying capacity models.
5. Advanced method by numerical modelling.

In standard fire resistance tests, only a very limited number of configurations can be tested due to high cost of testing, and the test specimens are limited to specific dimensions and boundary conditions due to the constraints of fire testing furnaces. Therefore, manufacturers of thin-walled steel construction systems always attempt to extend the range of their fire test results to different situations via expert opinion and engineering judgement. This may be possible under very limited circumstances (e.g. from a thinner section to a thicker section), and the extensions are within a small deviation from the tested configuration, for example, to extend the height of application by no more than 10%. Because this method is not based on a thorough investigation of heat transfer, load-bearing capacity and fire performance of the construction, it is by its nature very approximate and tends to be conservative in general, but can also be unsafe.

Both the limiting temperature method and the extension to ambient temperature design methods are covered in Eurocode EN 1993-1-2 (CEN 2005) and AS/NZS 4600 (SA 2018). The limiting temperature method simply states that the temperature in a class 4 (thin-walled) section should not exceed 350°C. This is a very conservative value in the majority of cases. Extending the ambient temperature design method consists of using ambient temperature effective widths for thin-walled steel structures (EN 1993-1-3 (CEN 2006a)/EN 1993-1-5 (CEN 2006b)) and replacing the ambient temperature properties of steel by those at elevated temperatures. The implicit assumption of this extension method is that the temperature distribution in the thin-walled (class 4) steel members is uniform. This assumption is not suitable in the majority of cases in which thin-walled steel members are protected by fire-rated boards and have steep non-uniform temperature distributions due to fire exposure from one side only. With non-uniform temperature distribution, it would not be sensible to assume that the temperature distribution is uniform, even at the maximum temperature. This is because this assumption could lead to very conservative answers in some cases and very unsafe solutions in other cases when the effects of thermal bowing play a significant role. Extending the ambient temperature method also assumes similar mechanical property characteristics at ambient and elevated temperatures, which is not true for cold-formed steels.

On the other hand, sophisticated finite element methods can deal with any type of thin-walled structures. However, using them requires deep and specialist knowledge as well as skills. They are best used as research tools rather than for everyday design.

This book will thus focus on simplified, but flexible, calculation methods. The simplified methods involve the following general steps:

1. Quantifying the fire exposure condition: For heat transfer and subsequent structural/ mechanical calculations, the fire exposure condition is presented in the form of fire temperature–time relationship. For simplicity and to cover the majority of common applications, only nominal fire temperature–time relationships will be used in this book.
2. Calculating the temperature field in the structural member under the above fire exposure: Thin-walled steel structures usually form part of panel construction (walls/floors) and are exposed to fire from one side. This makes the simple analytical solutions in Eurocode EN 1993-1-2 (CEN 2005) not relevant, because they are for uniformly heated members. This book will present an alternative simplified method.
3. Calculation of the remaining load-carrying capacity of the structural member at elevated temperatures and comparison with the applied load: The direct strength method, recently developed by Shahbazian and Wang (2014a), has been demonstrated to be suitable and relatively simple to implement. This will form the basis of the calculation method of this book. AS/NZS 4600 (SA 2018) has included simple methods in its appendix for thin-walled steel-framed walls and floors based on the research at Queensland University of Technology, which is an extension of the earlier work by Feng et al. (2003d).

1.4 SCOPE AND LAYOUT OF THIS BOOK

As has been mentioned previously, thin-walled steel construction involves many details to ensure fire integrity and to satisfy various other performance requirements such as sound and thermal insulation. An example is shown in Figure 1.3. It is assumed in this book that the integrity of panels is not compromised due to any issues with detailing, and hence the boards and interior insulation (if any) stay with the steel members throughout fire exposure.

In the majority of cases, thin-walled steel sections form part of walls and floors, and fire exposure is from one side of the construction. In this book, it is assumed that fire exposure is on the bottom surface of the floor and on the interior surface of the wall. The situation of fire exposure on the top surface of floors is unlikely to be as critical as on the bottom surface: the upward thermal bowing of the floor under fire exposure on the top is counteracted by the effects of gravity loading causing the floor to move downward. Similarly, if a

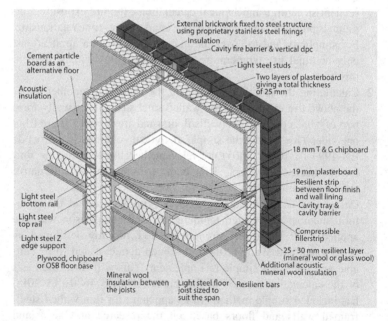

FIGURE 1.3 An example of detailing in thin-walled steel construction. (From Steel Construction Institute (SCI), Fire resistance of light steel framing, SCI Publication P424, Steel Construction Institute, Ascot, UK, 2019. With permission.)

wall is exposed to an external fire, as in the case of external fire spread, the external fire temperature will be much lower than the internal fire temperature. Therefore, the external fire exposure will in general be less detrimental to the structure than interior fire exposure.

Therefore, the general scope and assumptions of this book are:

- Simplified calculation methods to calculate temperatures and load-bearing capacities of thin-walled steel members.
- Fire from underneath floors.
- Fire inside fire compartment.
- No integrity failure.

When performing calculations to evaluate the fire resistance of structures, it is important to have accurate input data of temperature-dependent material properties. This is even more so for thin-walled steel structures because the elevated temperature properties of thin-walled steel (cold-formed) can vary within a large range, and the thermal properties of the non-load-bearing materials in thin-walled construction (gypsum plasterboard, interior insulations) are critically dependent on temperature.

Therefore, based on the aforementioned scope and important issues, this book will cover the following topics:

Chapter 2 will present applications of thin-walled steel structures. In particular, it will introduce different types of thin-walled steel structures, focusing on recent innovations.

Chapter 3 will present fire resistance tests of thin-walled steel construction. It will provide guidance on the performance of different types of construction and how detailing affects fire resistance/performance.

Chapter 4 will present various aspects of simplified modelling of fire resistance for thin-walled steel construction, including an introduction to fire behaviour and heat transfer modelling, guidance on the main features of heat transfer in thin-walled steel construction and how to simplify thin-walled steel construction for heat transfer modelling, and important features of modelling thin-walled steel structures at elevated temperatures.

Chapter 5 will assess elevated temperature mechanical properties of thin-walled (cold-formed) steels and present temperature-dependent thermal properties of non-load-bearing materials relevant to thin-walled steel construction.

Chapter 6 will present an assessment of different simplified calculation methods and present robust and accurate design calculation methods for thin-walled steel beams and columns at elevated temperatures. Suitable examples will be provided to demonstrate how the various simplified calculation methods can be implemented.

Applications of Thin-Walled Steel Structures

2

Applications of thin-walled steel structures are expanding rapidly owing to their many advantages as will be described in the next section. The increased applications are assisted by advances made in the field including the flexibility with which thin-walled steel structures can be made that lends to innovation of thin-walled steel sections and systems. For thin-walled steel structures used in the building sector, fire safety is a critical consideration. Before presenting detailed methodologies for quantifying the fire resistance of thin-walled steel structures, this chapter discusses applications of thin-walled steel structures in the building sector, recent advances and innovations, and introduces the key parameters that provide and enhance the fire resistance of thin-walled steel construction.

2.1 INTRODUCTION

Thin-walled steel structural components are made using a cold-forming process (roll-forming or press-braking) at ambient temperature. Unlike hot-rolled and welded heavy steel sections, which are only available in limited shapes and sizes, thin cold-formed steel sections can be custom designed and made to enhance structural and cost efficiencies and to suit specific applications. Unlipped and lipped channel sections (Figure 2.1) are still commonly used in the construction of wall, floor and roof systems. Z-sections are also used in applications such as purlins and joists. However, complex sections shown in Figure 2.2 can be easily roll-formed using low- and high-strength cold-rolled steel coils with thicknesses in the range of 0.55–6.35 mm since automated advanced manufacturing technologies are available now.

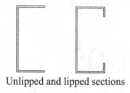

Unlipped and lipped sections

FIGURE 2.1   Simple channel sections.

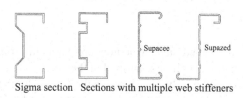

Sigma section   Sections with multiple web stiffeners

FIGURE 2.2   Complex thin-walled steel sections.

Thin-walled steel sections are not limited to non-load-bearing units such as partitions in buildings and are increasingly used in many primary load-bearing applications. They possess many beneficial characteristics such as high strength-to-weight ratio (lightweight), easier and faster fabrication and construction, enhanced durability through galvanizing (zinc or zincalume coating), termite-resistance, non-combustibility and sustainable features such as recyclability, reduced energy consumption and emissions, which have led to increased development and use of thin-walled steel construction systems. Most importantly, American, European and Australia/New Zealand cold-formed steel design standards, AISI S100 (AISI 2016), EC3 Part 1.3 (CEN 2006a) and AS/NZS 4600 (SA 2018), have been significantly enhanced in recent years through the adoption of many useful outcomes from research and development projects, allowing engineers to undertake economical and safe building designs. All of these have led to the expansion of thin-walled cold-formed steel construction from single storey residential buildings (Figure 2.3) to mid-rise residential and non-residential buildings of up to 12 storeys (Figure 2.4) in many countries around the world.

FIGURE 2.3 A residential building.

FIGURE 2.4 A mid-rise building. (Courtesy of Super Stud Building Products, Edison, NJ.)

2.2 RECENT INNOVATIONS AND ADVANCES IN COLD-FORMED STEEL INDUSTRY

Thin-walled steel sections and structural systems can be easily manufactured. Because of which, there have been many research and development studies to develop optimised thin-walled steel structural products and systems, which helped expand the applications of thin-walled steel structures.

2.2.1 Innovative Lightweight and Structurally Efficient Sections

Lipped channel sections (Figure 2.1) are commonly used as studs, joists and chord and web members in wall and floor panels and roof trusses and their dimensions, especially web height and thickness, are varied accordingly to suit the application. For example, 90 × 35 × 10 × 0.75 mm lipped channels are commonly used as wall studs while floor joists are commonly 180 × 40 × 15 × 1.15 mm lipped channels. Such thin-walled steel sections under compression or bending actions are subject to buckling modes such as local buckling, distortional buckling and flexural buckling (compression only) and flexural torsional buckling and their interactions. However, innovative sections can be developed to eliminate or delay one or more of the above-mentioned buckling modes to enhance their load-bearing capacities. For example, to eliminate local buckling effects, longitudinal web and flange stiffeners are used while complex return lips are used to enhance distortional buckling capacity.

Figure 2.2 shows two such sections (Supacee and Supazed) developed by the University of Sydney researchers in collaboration with an industry partner (Pham and Hancock 2013). Their research demonstrated the enhancements in bending capacity of these sections (up to 22% compared to conventional channel sections) and led to thinner and lightweight sections enabled also by the use of high-strength steels (G450-nominal yield strength of 450 MPa). These sections are currently being manufactured and distributed by Bluescope Lysaght in the Asia-Pacific region. Similar complex sections shown in Figure 2.5 have been researched extensively and used in different countries. In recent times, optimisation techniques such as simulated annealing algorithm with fabrication and geometric end-use constraints are being used to develop innovative and structurally efficient thin-walled steel sections for use in various applications (Leng et al. 2014).

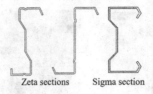

Zeta sections   Sigma section

FIGURE 2.5   Other complex sections.

In general, thin-walled cold-formed steel sections are open, monosymmetric or asymmetric sections, and are therefore subjected to more complicated buckling modes. To overcome the shortcomings associated with such sections, a series of hollow flange sections (HFS) were produced from a single strip of steel by an Australian steel product manufacturer using a patented manufacturing process based on simultaneous cold-forming and dual electric resistance welding. The first such section known as hollow flange beam (HFB) with torsionally rigid, triangular hollow flanges was a doubly symmetric section with no free edges and was used as flexural members (Figure 2.6). It eliminated local and distortional buckling effects and was shown to be 40% lighter than conventional cold-formed steel sections (Dempsey 1990). This was followed by a mono-symmetric hollow flange channel (HFC) section with rectangular flanges (Figure 2.6), which was produced to eliminate the connection problems associated with triangular flanges. Both sections were successfully used in Australia and New Zealand for many years in a range of applications. However, due to commercial reasons, their manufacturing operations in Australia were discontinued by 2014. Structural behaviour and moment and shear capacities of both these sections were investigated extensively using experimental and numerical studies by Queensland University of Technology researchers (Avery et al. 2000, Keerthan and Mahendran 2011, Anapayan and Mahendran 2012), which led to useful design capacity equations and tables that were used by their manufacturers and engineers.

Despite discontinuation of such structurally efficient HFS, there were industry expectations to replace them with alternative sections. Hence a new rivet-fastened HFC section (Figure 2.6) was developed, and their bending, shear and web crippling capacities were extensively researched by Queensland University of Technology researchers (Siahaan et al. 2016). This thin-walled steel section can be manufactured at a reduced cost in comparison with welded HFC and thus offers a cost-effective and safe solution to the building sector. Research on the use of HFS as wall studs and floor joists has demonstrated their efficiencies and enhancements over conventional sections (Jatheeshan and Mahendran 2016a, Kesawan and Mahendran 2017a).

HFB   HFC   Riveted HFC   MCO

FIGURE 2.6   Hollow flange sections.

There is great potential to use similar HFS sections and optimised C- and Z-sections in a range of applications including wall, floor and truss systems. Recently, an HFS similar to HFC (Figure 2.6) was developed and used in modular units in Korea due to their excellent performance and cost-efficiency (Ha et al. 2016). Using simultaneous cold-formed and electric welding process, the so-called MCO sections with 100–120 mm flange widths are being produced with depths in the range of 200–390 mm and thicknesses in the range of 4.5–10 mm. A smaller clinched HFS is also used as truss members in Australia.

2.2.2 Prefabricated Structural and Modular Units

The availability of new automated roll-forming facilities means that complex and optimised thin-walled steel sections can be easily cold-formed to produce them with tight tolerances and custom cut lengths. The new facilities can integrate CAD design to produce pre-cut, punched and sized steel members, which can then be readily used to make the basic structural wall and floor panels and roof trusses inside a factory in a controlled environment. Figure 2.7 shows typical prefabricated wall and floor panels with suitable studs and joists at 600 mm centres (also 300 and 400 mm). They can all be then transported to the site (Figure 2.7) and assembled using, in most cases, the efficient self-drilling screw fasteners in constructing low- and mid-rise buildings (Figures 2.3 and 2.4). This off-site manufacturing of light steel framing (wall and floor panels and roof trusses) will considerably reduce labour cost and waste in the factory and

FIGURE 2.7   Wall and floor panels and roof trusses.

on site, and accelerate construction process (30%–35% faster). The basic wall and floor panels can also be lined with the required linings and insulations in the factory and then transported to site. They can also be used to build modular units in the factory, and then transported and used to construct modular buildings (Figure 2.8), an increasing trend observed recently. Prefabricated walls are also used as non-load-bearing infill walls in multi-storey concrete- or steel-framed buildings. In North America, Europe and Australia, there are well-established manufacturers of thin-walled steel sections who are suppliers of both prefabricated (planar) light steel framing systems and modular (volumetric) units made of such sections for specific applications in the construction sector.

Light steel buildings built using prefabricated steel framing or modular units offer many advantages. The use of lightweight systems leads to significantly reduced foundation loads. When appropriately designed, they possess good performance characteristics in relation to fire resistance, thermal insulation and acoustic insulation. The fabrication and construction methods used offer significant reduction in production and site waste, construction safety benefits and improved quality of final products (Lawson and Way 2016). Load-bearing and non-load-bearing light steel framing can be used in low- and mid-rise buildings. Lateral stability is provided by walls that have in-built K-strap or flat sheet cross-bracing or board lining. For mid-rise buildings, larger stud sections or back-to-back or nested stud sections can be used in the lower

FIGURE 2.8 A modular building. (From Steel Construction Institute (SCI), Fire resistance of light steel framing, SCI Publication P424, Steel Construction Institute, Ascot, UK, 2019. With permission.)

storeys, and then both stud size and thickness (0.75–3 mm) can be reduced for the higher storeys. Light steel buildings can be easily modified and extended if required, disassembled and reused at another site.

2.3 FIRE RESISTANCE OF THIN-WALLED STEEL STRUCTURES AND METHODS OF ENHANCEMENT

Except for some components such as wooden floor boards and EPS insulation panels (if used as external cladding), all other components (cold-formed steel, boards and insulations) of light steel wall and floor panels are non-combustible. Hence, they do not contribute to the fire load of the building. However, the section factor (exposed surface area to volume ratio) of thin-walled steel wall studs and floor joists is high, and thus they heat up quickly in a fire unless

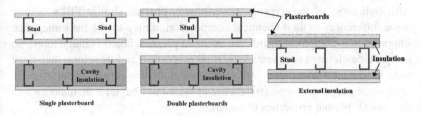

FIGURE 2.9 Protected thin-walled steel wall systems.

protected adequately. Depending on the application, thin-walled steel wall and floor systems may require fire resistance levels or fire resistance ratings up to four hours under the three criteria of structural adequacy/stability, integrity and insulation. For consistency, this book uses the term fire resistance level (FRL) to refer to the period of standard fire exposure under which a fire-resistant construction should maintain its fire-resistant function. Different names are used in different standards, and other names include fire resistance period and fire resistance rating. Non-load-bearing walls require only FRLs under the integrity and insulation criteria. In order to provide the required FRLs, thin-walled steel members are always protected by fire-rated boards such as gypsum-based boards and calcium silicate boards. Some examples of board protection are shown in Figure 2.9. Fire resistance depends on several key parameters, and this section explains their effects briefly and discusses methods to provide enhanced FRLs. Chapter 3 will provide more details of fire resistance/performance of thin-walled steel structures.

2.3.1 Fire Protective Boards

Fire resistance of thin-walled steel wall and floor systems depends significantly on fire protective boards used as linings and their arrangement. These boards protect thin-walled members from direct fire exposure and delay the temperature rise in steel. The time-temperature profile developed in steel studs/joists and across the depth of the wall/floor panel in fire depends on the thermal properties, fall-off temperature and joints of the fire protective boards. The time-temperature profiles in studs/joists directly influence the structural resistance in fire while the temperature profile on the unexposed side determines the insulation performance. Therefore, the elevated temperature thermal properties of fire protective boards, especially specific heat, relative density (mass loss) and thermal conductivity, are critical. However, they vary among them due to the differences in their chemical compositions and manufacturing processes.

Although most of the gypsum plasterboards provide similar FRLs despite some differences in their chemical composition, it may not be the same when imported non-compliant boards are used. At present, board manufacturers do not provide elevated temperature thermal properties, and therefore, they need to be determined first, particularly if thermal modelling is to be undertaken for fire design purposes. Some guidance is provided in Chapter 5 on temperature-dependent thermal properties of materials.

Unlike gypsum plasterboard that provides consistent fire protection to steel wall and floor systems, the use of other types of boards such as MgO board and PCM (phase change material) board often reduces the FRLs of light steel walls even though they provide better impact resistance, thermal and acoustic insulation. Recent fire tests of MgO board-lined walls (Rusthi et al. 2017) showed premature integrity failure within 30 min. This is due to board joint opening and board cracking caused by significant mass loss with increasing temperature (about 45% compared to 20% in gypsum plasterboard). Hence it is clear that elevated temperature thermal property tests of boards and/or fire tests of walls/floors should be undertaken before they are used.

The FRLs of wall and floor systems can be increased either by enhancing the specific heat and relative density values of gypsum plasterboards or by reducing their thermal conductivity values. Baux et al. (2008), Baspinar and Kahraman (2011) and Keerthan et al. (2013) have shown that elevated temperature properties of plasterboard can be enhanced by adding chemical fillers and additives. Further research is needed to develop new boards with desirable thermal properties to provide considerably higher FRLs to wall and floor systems.

Gypsum plasterboards crack and fall-off due to dehydration and calcination reactions, which leads to sudden localised temperature rise in steel studs/joists and early failures in fire (Gunalan et al. 2013). Enhanced plasterboards with reduced cracking and fall-off at elevated temperatures would increase the fire resistance of light steel walls.

Board joints are unavoidable in thin-walled steel wall and floor systems (Figure 2.10). They are staggered so that the same stud/joist does not have a joint along the length on both flanges and are located in different directions when multiple boards are used. Although joints are fully sealed, they open up when exposed directly to fire and allow localised temperature rise, leading to reduced FRLs as observed in many full-scale fire tests (Gunalan et al. 2013). This effect is significant in single board lined walls with joints along the steel members. Good detailing and workmanship can eliminate the weakness caused by board joints to some extent. Kesawan and Mahendran (2017b) recommended improved joint methods (Figure 2.11) to eliminate the detrimental effects of joints; however, they may be considered labour intensive. The new SCI guide (SCI 2019) provides many examples of good detailing practice adopted by UK manufacturers of thin-walled steel structures.

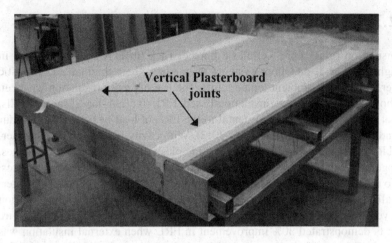

FIGURE 2.10 Plasterboard joints. (With kind permission from Springer Science+Business Media: *Fire Technol.*, A review of parameters influencing the fire performance of light gauge steel framed walls, 2017, Kesawan, S. and Mahendran, M.)

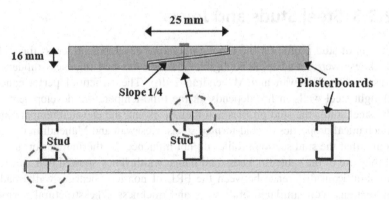

FIGURE 2.11 Improved board joints. (With kind permission from Springer Science+Business Media: *Fire Technol.*, A review of parameters influencing the fire performance of light gauge steel-framed walls, 2017, Kesawan, S. and Mahendran, M.)

2.3.2 Insulation

Insulation (glass fibre, cellulosic fibre or rock fibre) is commonly used in light steel wall and floor systems to enhance their thermal and acoustic performance at ambient temperature. The use of cavity insulation has been shown to increase the insulation-based FRL for non-load-bearing walls, but it was found to be detrimental to the FRL of load-bearing walls (Kodur and Sultan 2006, Gunalan et al. 2013). With cavity insulation, the considerable rise in temperatures on the fireside plasterboards and stud hot flanges, together with a larger temperature gradient, led to premature stud failures. To overcome this problem, Kolarkar and Mahendran (2008) proposed externally insulated light gauge steel-framed (LSF) walls, where an insulation layer was sandwiched between two plasterboard layers (Figure 2.9). Fire tests demonstrated 30% improvement in FRL when external insulation was used. The use of similar external insulation systems will enhance the fire resistance of both non-load-bearing and load-bearing walls. If composite panels consisting of boards and insulation can be prefabricated and installed, the installation cost will remain the same. Enhanced insulation materials with improved thermal properties can also be developed by varying their compositions.

2.3.3 Steel Studs and Joists

Effects of stud profile on the time-temperature development across the wall thickness were found to be negligible although the stud thickness influence was visible (Kesawan and Mahendran 2016). The structural performance of light steel walls in fire depends on the time-temperature development in the steel studs, the stud profiles and their sizes and the elevated temperature mechanical properties of cold-formed steels. Kesawan and Mahendran (2016) found that the stud section profiles do not influence the thermal performance if they are of the same thickness and flange width. They showed that there is no significant difference between the FRL of non-load-bearing walls made of sections with similar overall sizes and thickness. The structural performance of LSF walls in fire can depend on the thermal performance (temperature development) within the wall, steel sections (profiles and sizes) used and elevated temperature mechanical property reduction factors of cold-formed steels. However, fire tests and numerical studies (Kesawan and Mahendran 2016, 2017a) have shown that wall and floor systems made of studs/joists with different profiles have the same FRL for a given load ratio if their depth and flange widths are the same. Increase in stud thickness is likely to improve the FRL of LSF walls. Effects of web depth on the FRL of walls are dependent on the type of failure mode of studs (section yielding or local buckling or major axis buckling) and thermal bowing deflections. In summary, for a chosen wall/floor depth, FRL cannot be improved by stud/joist section profile. However, elevated temperature mechanical property reduction factors vary depending on the type/grade of steel used. Ariyanayagam and Mahendran (2018) showed that the use of steels of varying strength grades significantly influenced the FRL of load-bearing walls. Hence developing and using cold-formed steels with higher elevated temperature mechanical property reduction factors, especially in the range of 400°C–700°C, can provide higher FRLs for thin-walled steel wall and floor systems.

2.3.4 Steel Sheathing

Thin-walled steel stud walls lined with steel sheathing are increasingly used in seismic regions. The use of steel sheathing together with plasterboard sheeting provides higher FRLs for non-load-bearing systems because it restricts plasterboard fall-off and retains vaporized water. However, its effect on the FRL of load-bearing walls was found to be minimal for a given load ratio (Dias and Mahendran 2019a), in which a structurally efficient web-stiffened section was used by eliminating local and distortional buckling effects. This pushed up the critical mode of failure to be major axis flexural buckling and the use of steel sheathing gave an increased wall capacity at ambient temperature and also under fire conditions without compromising the FRL (similar FRL for a given load ratio). Hence, choosing structurally efficient stud/joist profiles in combination with sheathing can provide enhanced FRLs.

2.4 SUMMARY

The last section has briefly explained the effects of key parameters on the fire resistance of thin-walled steel construction systems and suggested methods to enhance fire resistance. The next chapter will provide further details of fire performance characteristics of different types of thin-walled construction systems based on fire resistance tests.

Fire Resistance Tests

3

Because of complexity of detailing of thin-walled steel structural systems, most fire design standards worldwide use the so-called standard fire tests to determine their fire resistance level (FRL). Standard fire resistance tests are also used by researchers to determine and investigate the FRL of the constructed system as a function of key influential parameters. This chapter first presents the standard fire resistance test in general, and then describes the fire resistance characteristics of different types of thin-walled steel wall and floor systems.

3.1 STANDARD FIRE RESISTANCE TEST

When exposed to fire, thin-walled steel structural elements and assemblies heat up quickly, due to their high section factor, and lose their strength and stiffness rapidly. Hence, they are normally protected by fire-rated boards to delay the temperature rise. For example, light gauge steel-framed (LSF) wall systems made of lipped channel section studs as vertical members and unlipped channel section tracks as top and bottom horizontal members are lined with single or double layers of fire rated boards such as gypsum plasterboard, calcium silicate board, fibre cement board or other equivalent boards of thicknesses in the range of 6–25 mm and with or without cavity or external insulation using rock fibre, glass fibre or cellulose fibre in different thicknesses and densities (Figure 3.1a). Depending on whether they are used as external or internal walls or load-bearing or non-load-bearing (partitions only) walls, such arrangements and sizes will vary significantly among them. Noggings and bracings are also provided for lateral resistance and construction purposes. LSF floor systems are similar to LSF wall systems, however, larger lipped channel or Z-sections are used with fire rated boards attached to the ceiling side (bottom surface) and plywood or particle board on the unexposed

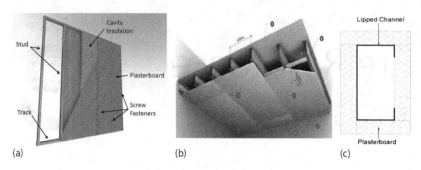

FIGURE 3.1 Protected thin-walled steel structural elements and assemblies. (a) LSF wall system, (b) LSF floor system (Promatect), (c) protected beam/column. (Courtesy of Promatech, Moorestown, NJ.)

side (top surface) (Figure 3.1b). Similarly, thin-walled steel structural elements such as beams and columns can also be protected (box-protection) using fire rated boards as shown in Figure 3.1c.

Fire resistance of thin-walled steel structural elements and wall and floor systems depends on many parameters such as the type of steel and thickness, member size, type of board and insulation and their thickness and density, wall and floor configuration and load ratio. Different countries and regions have their own standard fire resistance test standards; however, they are all very similar. In standard fire resistance tests of LSF systems, full size walls/floors (minimum size of 3 m × 3 m for wall panels and 3 m × 4 m for floor panels) are exposed to the standard fire time-temperature curve (Figure 3.2) on one side using a gas or oil furnace. If they are load-bearing, they are subjected to a pre-determined load based on the chosen load ratio during the entire fire test. The load ratio is defined as the applied load in a standard fire test divided by the ambient temperature capacity. Full size panels are used in order to include the effects of construction method used (board joints, connections, etc), thermal expansion, shrinkage, cracking, fall-off, ablation and localized deformations and damage.

The origins of the standard fire time-temperature curve date back as far as 1903 based on wood fuel-burning furnaces. It is considered to simulate a fully developed compartment fire. The international standard ISO 834 test fire curve (ISO 1999) is defined by Eq. (3.1).

$$T_f = 20 + 345 \log 10 \left(8\,t + 1\right) \tag{3.1}$$

where:

T_f is the average furnace temperature (°C) at time t

t is the time elapsed (min)

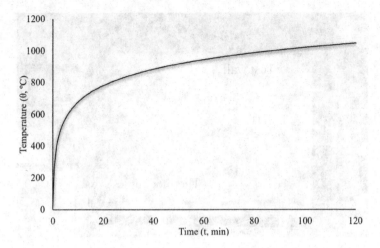

FIGURE 3.2   Standard fire time-temperature curve. (From ISO 834-1, Fire resistance tests – elements of building construction, Part 1: General requirements. International Organization for Standardization, Geneva, Switzerland, 1999.)

Depending on the intended function of the tested system, the fire resistance level (FRL) of the tested system is the time when one of the following three fire resistance criteria is reached:

- Load-bearing – Structural failure due to rapid loss of load-bearing capacity or deflection/deflection rate exceeding their corresponding limits.
- Integrity – Passage of hot gases/flames through the construction assembly due to crack formation. Integrity failure is determined by using a cotton pad and gap gauges.
- Insulation – The unexposed side temperature exceeds the ambient temperature by 140°C on average or by 180°C at any point on the unexposed side.

Figure 3.3 shows a typical fire test set-up used to assess a cold-formed LSF wall system where pre-determined loads are applied to the studs via the bottom track. Fire tests of LSF floor systems are conducted in a horizontal furnace with fire exposure from underneath after applying the pre-determined vertical loads to the floor boards.

Fire testing using the standard fire time-temperature curve gives good comparative results for building components tested under identical conditions, and also valuable basic test data. They are commonly used by product

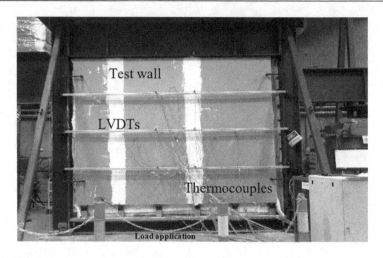

FIGURE 3.3   Fire test set-up for LSF wall systems.

manufacturers, industry associations and some building codes to develop generic or proprietary FRLs for a range of thin-walled steel wall and floor systems. Although there is growing criticism that the standard fire time-temperature curve does not represent fires in real situations, standard fire tests are used regularly by researchers worldwide to investigate the effects of key parameters on the fire resistance of thin-walled steel construction systems, to compare their FRLs and to develop new construction systems with enhanced fire resistance. The next three sections will discuss the fire resistance tests of different types of thin-walled steel construction systems (walls, floors and members) and their fire performance characteristics. Although product manu-facturers have carried out numerous standard fire resistance tests, their results are not publicly available. Therefore, the fire resistance test results in this book are sourced from publicly available research literature.

3.2 FIRE RESISTANCE TESTS OF WALLS

3.2.1 General Behaviour of LSF Walls in Fire

Gerlich et al. (1996) conducted three standard fire tests of channel section stud walls vertically lined with one layer of gypsum plasterboard on both sides under combined axial loading and bending. Studs were made of 1.15 mm

Grade 300 steel (2.85 m height) and 1.0 mm Grade 450 steel (3.6 m height) and used at 600 mm spacing with a central row of noggings without any interior insulation. The tests were conducted on horizontally placed panels by loading via the tracks. Major axis flexural buckling associated with local buckling of mid-height cold flanges occurred in two tests while in the third test flexural torsional buckling occurred due to the loss of lateral restraint to the compression flange by the thinner (9.5 mm only) ambient side plasterboards. With the use of 12.5 and 9.5 mm boards, 32 min fire resistance was reached while using 16 mm boards gave 72 min fire resistance.

Kodur and Sultan's (2006) experimental study included 14 LSF wall panels with 0.84 and 0.912 mm thick studs spaced at 406 mm (610 mm in one test) and lined with one or two 12.7 mm gypsum boards on both sides (15.9 mm boards in two tests). Glass fibre, rock fibre and cellulose fibre cavity insulations were used. Local buckling of studs was the dominant structural failure mode except in two tests where overall buckling was observed. Fire resistance was reduced when cavity insulation was used, with cellulose fibre performing better than others.

Alfawakhiri (2001) conducted six standard fire resistance tests of load-bearing LSF walls made of 0.912 mm thick 90 mm studs (low yield strength of 228 MPa). Major axis flexural buckling of studs occurred towards the furnace with compressive failure of the cold flange near mid-height. When cavity insulation was used, stud hot flange temperatures were increased while cold flange temperatures were reduced, and the wall moved away from the furnace and failed by hot flange compression failure. Using cavity insulation reduced fire resistance. The use of noggings and ambient side cross-bracing was considered to have eliminated flexural-torsional and minor axis flexural buckling of the studs. Local compressive failures of studs occurred at one of the four holes located along the length. The use of resilient channels slightly reduced the fire resistance times.

Feng and Wang (2005) conducted six standard fire tests of load-bearing LSF wall panels of 2.2 × 2.0 m lined with 12.5 mm gypsum board lining and cavity insulation. Lipped channel studs of 100 × 54 × 15 mm made of 1.2 and 2.0 mm thick S350 steels were used with two holes near the ends. Increasing the load ratio (0.2, 0.4 and 0.7) decreased the stud failure time significantly. Except for one test (local buckling at the hole), the studs failed by flexural torsional buckling. This was considered to be due to the inability of fireside plasterboard to restrain the studs although the ambient side plasterboard was able to prevent minor axis flexural buckling. Their study noted that plasterboard fall-off could have been triggered by stud failure, instead of plasterboard fall-off causing stud failure.

Zhao et al. (2005) conducted 29 tests of 1.2 × 2.8 m LSF walls with two studs spaced at 600 mm with gypsum plasterboards on one or both sides, and with or without cavity insulation. Studs of varying sizes were used and loaded concentrically or eccentrically for load ratios from 0.2 to 0.6. Test results showed that stud failure is essentially governed by the maximum temperature in fire. Steel studs supported by plasterboards on both sides failed by major axis flexural buckling with local buckling of the cold flange. This failure mode was observed with eccentric loading towards the fireside. In some other tests, the same failure mode occurred but with movement away from the fire and compressive failure of the hot flange. These observations show the presence of complex stud failure modes due to combined thermal and structural effects and are similar to those observed by other researchers as discussed above.

Gunalan et al. (2013) conducted 11 full-scale fire tests of load-bearing walls made of 1.15 mm high-strength G500 cold-formed steels. Single or double gypsum plasterboard lining of 16 mm thickness was used with or without rock fibre, glass fibre and cellulosic fibre cavity insulation. Typical plasterboard and stud time-temperature curves from their standard fire tests of double plasterboard–lined walls are shown in Figure 3.4. They are similar to those observed by earlier researchers and show that gypsum plasterboards are able to keep the cavity temperatures low and delay steel stud temperature rise for a long time. The figure also shows presence of non-uniform temperature distributions (temperature gradient) across the wall/stud depth. Many researchers (Kaitila 2002, Feng et al. 2003a, Gunalan et al. 2013) have confirmed that it is acceptable to assume uniform temperatures in the flanges and lips and a linear web temperature distribution as shown in Figure 3.5.

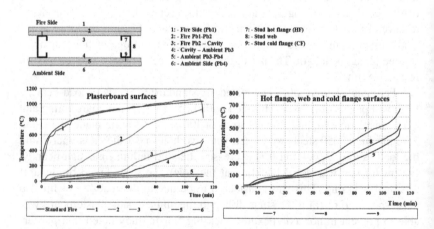

FIGURE 3.4 Time-temperature curves from a standard fire test of LSF wall. (Reprinted from *Thin-Walled Struct.*, 65, Gunalan, S. et al., Experimental study of cold-formed steel wall systems under fire conditions, 72–92, Copyright 2013, with permission from Elsevier.)

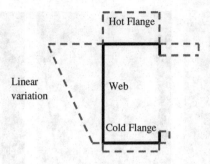

FIGURE 3.5 Simplified stud temperature distribution.

The development of non-uniform temperature distributions across the stud depth causes thermal bowing towards the fireside as illustrated in Figure 3.6 (original centroid X moves to Y at mid-height: $e\Delta_T$) while the centre of resistance of the stud shifts towards the ambient side due to the loss of stiffness of the hotter side of the steel stud, giving a neutral axis shift ($e\Delta_E$). This results in a net eccentricity of $e = e\Delta_T - e\Delta_E$. Hence the wall studs are subjected to a combined loading of axial compression and bending moment in fire. The net loading eccentricity (e) in Figure 3.6 will vary depending on the values of $e\Delta_T$ and $e\Delta_E$, which depend on the temperature distributions of the stud cross-section. The net e value is normally positive and hence the wall failure occurs by movement towards the furnace. However, in some cases, the wall can fail by a sudden movement away from the furnace near the end (Figure 3.7) depending on the net e value and the relative yield strengths of hot and cold flanges near the failure time as explained in Gunalan et al. (2013).

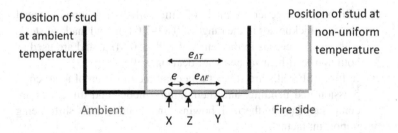

FIGURE 3.6 Effective centroid of studs in fire. (Reprinted from *Thin-Walled Struct.*, 65, Gunalan, S. et al., Experimental study of load bearing cold-formed steel wall systems under fire conditions, 72–92, Copyright 2013, with permission from Elsevier.)

FIGURE 3.7   LSF wall panels after failure. (Reprinted from *Thin-Walled Struct.*, 65, Gunalan, S. et al., Experimental study of load bearing cold-formed steel wall systems under fire conditions, 72–92, Copyright 2013, with permission from Elsevier.)

Gunalan et al.'s (2013) fire tests demonstrated integrity of the inner plasterboard until failure (Figure 3.7) and showed that there was sufficient lateral restraint to the studs from the plasterboards until failure. In fact, in all the fire tests of load-bearing walls with load ratios of 0.2 and above, steel stud failure occurred before integrity or insulation failure. This is despite plasterboard fall-off that started to occur when the inner plasterboard surface temperature reached about 900°C. Other researchers (Alfawakhiri 2001, Feng et al. 2003b, 2003c, Zhao et al. 2005) also suggested that it would be appropriate to assume sufficient plasterboard lateral restraints to prevent minor axis flexural buckling and flexural-torsional buckling failures of the steel studs.

From the above standard fire tests of load-bearing LSF walls conducted in many different countries, the following useful general observations are made:

- LSF walls are generally made of thin-walled steel-lipped channel studs of thicknesses in the range of 0.84–2.0 mm and made of steels with yield strengths in the range of 228–550 MPa, and are used as both non-load-bearing and load-bearing walls.
- In fire, LSF walls are subject to a combined loading of axial compression and bending, and their failure modes and directions are complicated with thermal bowing and neutral axis shift being important factors.
- The number, thickness and type of plasterboard lining are influential factors for the FRL of LSF walls. Protecting them with one or two layers of 9.5, 12.5 and 16 mm fire rated gypsum plasterboard has been successful in providing FRLs up to 120 min.

- Stud failure is governed mostly by the hot flange temperature reached in fire. Stud depth and thickness and the type of steel used can influence the fire resistance of walls.
- LSF wall failure is usually due to local buckling of the stud element but can also be caused by major axis flexural buckling or a combination of local and major axis flexural buckling (Figure 3.7). A single layer of 9.5 mm plasterboard was found to be unable to provide minor axis and twist restraints to the studs. Double layers of plasterboard were able to provide sufficient lateral restraint to the studs until failure. The use of noggings and bracing will provide adequate restraint against minor axis and flexural-torsional buckling.
- Plasterboard fall-off can allow heat inside cavity, leading to earlier failures and reduce FRLs of LSF walls. Its detrimental effects are more critical for a single layer of plasterboard-lined walls.
- Glass fibre, rock fibre and cellulosic insulation materials are commonly used as cavity insulation to provide sound insulation, thermal comfort and energy savings. However, they reduce the FRL of load-bearing walls.
- In all the fire tests of load-bearing walls with load ratios of 0.2 and above, stud load-bearing failure occurred before integrity or insulation failure. Increasing load ratios led to lower FRLs but the time-temperature curves are not altered by the load ratios.

Insulation and integrity-based fire resistance of non-load-bearing LSF walls was investigated by Feng et al. (2003c), Kolarkar and Mahendran (2012) and Ariyanayagam and Mahendran (2018) using both small-scale (300 × 300 mm to 1280 × 1015 mm) and full-scale (3 × 3 m) standard fire tests. Several numerical studies based on small-scale tests were also undertaken by Feng et al. (2003c) and Sultan (1996). Although small-scale fire tests cannot capture many important effects of full-scale fire tests, such as thermal bowing, large lateral deflections, cracking, ablation, fall-off of plasterboards and integrity failures, they are adequate in providing comparative results and the general trend. These studies showed that insulation-based FRL was improved by more than 15 min for 92 × 1.15 mm stud walls lined with one layer of 16 mm plasterboard when glass fibre insulation was used, by more than 30 min when the plasterboard thickness was increased from 13 to 16 mm, and by more than 100 min when two layers of 16 mm plasterboard were used. Fire tests of 3 m high walls showed that integrity failure did not occur when gypsum plasterboard linings were used. Although plasterboard joints play an important role in affecting the FRLs of load-bearing walls (see Section 3.2.5), their effects on the FRLs of non-load-bearing walls appear to be small.

The following subsections will discuss the effects of different influential factors.

3.2.2 Effects of Cavity Insulation on Fire Resistance

Ariyanayagam and Mahendran (2019) investigated the effect of cavity insulation on non-load-bearing and load-bearing of 92 mm stud walls. Four fire tests of 3 × 3 m walls and parametric finite element studies were conducted on LSF walls lined with one 16 mm gypsum board with and without glass fibre cavity insulation. Cavity insulation prevents heat transfer across the cavity by radiation and convection, thereby increasing temperatures of the fireside plasterboard and hot flanges while decreasing temperatures of the ambient side plasterboard surface and cold flanges, resulting in larger temperature gradients across the cavity and the stud than in uninsulated walls (Figure 3.8a). This led to higher thermal bowing deformations, and together with higher hot flange temperatures, resulted in considerably lower FRLs (stud failure times) of cavity-insulated LSF walls as shown in Figure 3.8b. Alfawakhiri et al. (1999), Kodur and Sultan (2006) and Gunalan et al. (2013) also confirmed the detrimental effects of cavity insulation on the FRL of load-bearing walls.

When cavity insulation becomes ineffective, the differences in temperatures between LSF walls with and without cavity insulation decrease, and therefore the adverse effect of cavity insulation on FRLs of load-bearing LSF walls diminishes. For example, the results in Figure 3.8a show that after glass

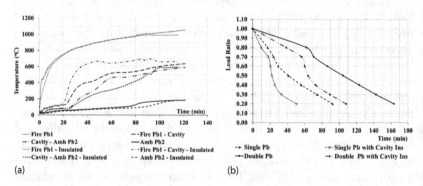

(a) (b)

FIGURE 3.8 Fire test results of cavity-insulated LSF walls: (a) time-temperature curves of plasterboard surfaces and (b) load ratio versus failure time curves. (Reprinted from *Constr. Build. Mater.*, 203, Ariyanayagam, A. and Mahendran, M., Influence of cavity insulation on the fire resistance of light gauge steel framed walls, 687–710, Copyright 2019, with permission from Elsevier.)

fibre becomes ineffective at about 650°C, the temperature differences between insulated and uninsulated walls started to decrease. Kodur and Sultan's (2006) fire tests showed that when cellulosic insulation was used, the load-bearing wall failed at 71 min, whereas the same wall with rock fibre insulation failed at 59 min. This is due to the difference in temperatures at which these cavity insulation materials become ineffective.

Load-bearing steel stud walls fail when the hot flange temperature is in the range of 400°C–600°C for common load ratios of 0.4–0.7. Hence, if the temperature at which cavity insulation becomes ineffective is below 300°C, it will not reduce the FRL of load-bearing LSF walls. Since cavity insulation is needed for other purposes such as sound insulation, thermal comfort and energy savings, such an approach should be pursued.

On the other hand, the use of cavity insulation reduced the ambient surface temperatures and thus increased the insulation-based FRL (Figure 3.8a). This implies that for non-load-bearing walls the use of rock fibre insulation with a higher effective temperature will enhance their FRLs. However, higher thermal deformations leading to large mid-height lateral deflections can trigger premature fall-off of already softened gypsum plasterboard on the fireside and thus reduce insulation-based FRLs.

3.2.3 Effects of External Insulation on Fire Resistance

To eliminate the large temperature gradient in LSF walls with cavity insulation which causes reductions to the FRLs of load-bearing LSF walls while still meeting the requirements for acoustic performance, thermal comfort and energy conservation, Kolarkar and Mahendran (2008) proposed the use of external insulation sandwiched between two plasterboards as shown in Figure 3.9a.

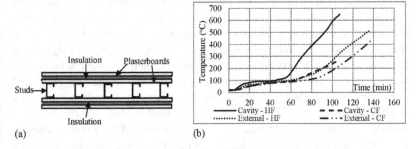

FIGURE 3.9 Composite panel: (a) with external insulation and (b) stud time-temperature curves (HF-hot flange, CF-cold flange).

The fire performance of the new LSF wall system was investigated using small-scale and full-scale fire tests and numerical studies by Kolarkar and Mahendran (2012), Gunalan et al. (2013), Kesawan and Mahendran (2015) and Chen and Ye (2014). The use of external insulation resulted in reduced temperatures of the inner plasterboard layer, and so its calcination and deterioration were delayed. It also allowed direct heat transfer from the fireside plasterboard to the ambient side plasterboard. Hence in contrast to cavity-insulated walls (Figure 3.8a), the stud hot flange temperatures and associated temperature gradients were significantly reduced as shown in Figure 3.9b. Fire tests showed that the FRLs of load-bearing lipped channel stud walls increased by about 25%, with rock fibre insulation providing the best outcome. Kesawan and Mahendran (2015) showed that similar improvements can be obtained for LSF walls made of other stud sections such as hollow flange channels. In terms of insulation performance, the corresponding FRLs are likely to be the same for both cavity and externally insulated panels based on both test and numerical results.

Furthermore, Gunalan et al.'s (2013) test results showed that the vertical plasterboard joints on the inner boards were protected by the external insulation layer even after the outer layer fall-off, thus preventing localised temperature rise in the stud hot flange. Therefore, plasterboard joints in LSF walls with external insulation are not as detrimental as those in cavity-insulated or uninsulated walls.

3.2.4 Effects of New Stud Sections on Fire Resistance

Kesawan and Mahendran (2015) investigated the possibility of increasing the fire resistance of LSF walls by using a new welded hollow flange channel (HFC) section of dimensions 150 x 45 x 15 x 1.6 mm as studs as shown in Figure 3.10. When the results of five fire tests of wall panels lined with two layers of 16 mm plasterboard were compared with those from Gunalan et al. (2013) who used conventional lipped channels, there were significant improvements to fire resistance times. However, their subsequent numerical studies showed that such improvements were not due to the new HFC stud geometry. Instead, they benefited from the enhanced elevated temperature mechanical properties of HFC due to the simultaneous cold-forming and electric resistance welding fabrication process used. These results demonstrate the importance of elevated temperature mechanical properties and how they may be improved to enhance fire resistance.

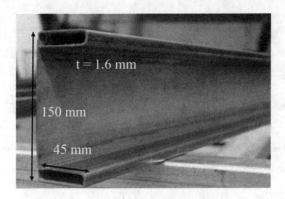

FIGURE 3.10   HFC stud sections. (Reprinted from *Thin-Walled Struct.*, 98(A), Kesawan, S. and Mahendran, M., Predicting the performance of LSF walls made of hollow flange sections in fire, 111–126, Copyright 2019, with permission from Elsevier.)

3.2.5 Effects of Plasterboard Joints on Fire Resistance

Plasterboard joints are located along the studs (vertically) in single plasterboard–lined walls. Although the recessed edges of the plasterboards are filled with joint fillers and sealed, plasterboards become soft, brittle and shrink and cause joint opening-up (Figure 3.11) in fire, especially due to their dehydration reactions (free and chemically bound water evaporates), and joints also become weaker and crack, and split due to short edge distances (10 mm). Such opening-up of joints on the fire side exposes the studs to direct fire attack and leads to rapid rise in the hot flange temperature of studs (Figure 3.11) and accelerate their structural failure (Ariyanayagam et al. 2016). Although the use of horizontal plasterboard joints along noggings could reduce such adverse effects, vertical joints along the studs still have to be used for longer walls. Other studies by Chen et al. (2012, 2013a, 2013b) have also shown the detrimental effects of plasterboard joints on fire resistance of LSF walls.

To mitigate against this shortcoming, Ariyanayagam et al. (2016) suggested a back-blocking joint arrangement in which the joints are located between the studs with 150 mm wide plasterboards as back-blocks (Figure 3.12). They showed that such joint arrangements increased the failure time of single plasterboard-lined load-bearing walls by about 25% as the stud hot flanges

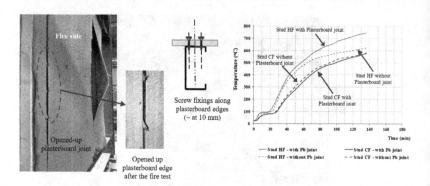

FIGURE 3.11 Cracking and opening-up of joints and their effects on stud time-temperature curves. (Reprinted from Ariyanayagam, A.D. et al., *Thin-Walled Struct.*, 107, Detrimental effects of plasterboard joints on the fire resistance of light gauge steel frame walls, 597–611, Copyright 2016, with permission from Elsevier.)

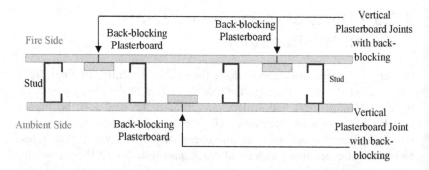

FIGURE 3.12 Plasterboard joints with back-blocks. (Reprinted from Ariyanayagam, A.D. et al., *Thin-Walled Struct.*, 107, Detrimental effects of plasterboard joints on the fire resistance of light gauge steel frame walls, 597–611, Copyright 2016, with permission from Elsevier.)

were not directly exposed to fire through the opened-up joints. They also showed that since the joint effect is localized by only affecting the stud hot flange and fireside plasterboard temperatures, its effect on the insulation-based FRL of non-load-bearing walls is negligible. Innovative plasterboard joint arrangements can also be used along the studs instead of back-blocking to reduce construction time/cost.

3.2.6 Effects of Other Types of Boards on Fire Resistance

Other types of fire protective boards are also being used in LSF walls, for example, calcium silicate boards, because of their improved properties relating to impact and moisture resistance. Calcium silicate boards are manufactured using calcium silicate hydroxide $(Ca_6Si_6O_{17}(OH)_2)$ also known as Xonotlite, mineral binder and additives.

Chen et al. (2012) investigated the use of two layers of 12 mm thick calcium silicate boards, and their fire tests showed explosive spalling at high temperatures, resulting in lower FRLs (58 min integrity failure) compared with gypsum plasterboard–lined walls. When a combination of 12 mm calcium silicate and 12.5 mm gypsum boards was used with the calcium silicate board on the inner side, FRL was improved to 92 min as the integrity failure associated with cracking was eliminated. Wang et al. (2015) also conducted full-scale tests of LSF walls lined with 9 mm calcium silicate board and observed cracking and spalling of the boards. Ariyanayagam and Mahendran (2017) conducted full-scale fire tests of non-load-bearing LSF walls with one 20 mm calcium silicate board lining in one test and then with a combination of 20 mm calcium silicate board and 16 mm gypsum plasterboard in the second test. No integrity failure associated with large cracks, spalling and fall-off occurred in their tests with an insulation-based failure after 130 min (Figure 3.13) compared to 94 min obtained for the 16 mm plasterboard-lined wall. In the second test, the average ambient surface temperature was 79°C even after 250 min, in comparison with 197 min for LSF walls lined with two layers of 16 mm plasterboard. The higher thickness of calcium silicate board helped to reduce the temperatures on the unexposed side and prevent integrity failure. The enhanced performance also benefited from glossy and smooth surface of calcium silicate boards (which have lower emissivity) and the presence of fibrous material throughout the calcium silicate board thickness.

Magnesium oxide board is another board that is being used as fire protection for LSF walls. Its main constituents are magnesium oxide (MgO) (40%–55%) and magnesium chloride $(MgCl_2)$ (20%–35%) with small percentages of perlite, woodchip and fibreglass. Thermal property tests of MgO boards by Rusthi et al. (2017) showed that they have almost 40%–50% mass loss during their dehydration reactions, compared to 20% for gypsum plasterboard (Figure 3.14).

Rusthi et al. (2017) conducted three full-scale tests of non-load-bearing LSF walls lined with one layer of two types of 10 mm MgO boards with differing percentages of MgO and $MgCl_2$. The first test panel had noggings

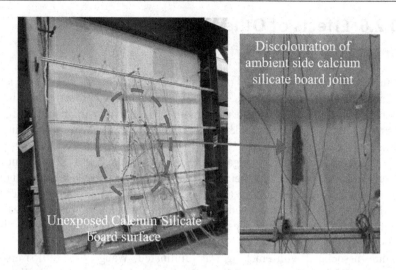

FIGURE 3.13 Calcium silicate board–lined wall. (Reprinted from *Thin-Walled Struct.*, 115, Ariyanayagam, A.D. and Mahendran, M., Fire tests of non-load bearing light gauge steel frame walls lined with calcium silicate boards and gypsum plasterboards, 86–99, Copyright 2017, with permission from Elsevier.)

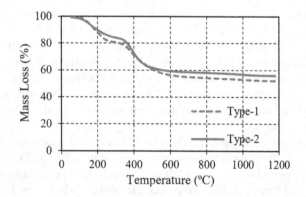

FIGURE 3.14 Mass loss of MgO boards. (Reprinted from *Fire Saf. J.*, 90, Rusthi, M. et al., Fire tests of magnesium oxide board lined light gauge steel frame wall systems, 15–27, Copyright 2017, with permission from Elsevier.)

at 600 mm spacing while the third one had glass fibre cavity insulation. The cavity-insulated wall panel exhibited higher hot flange temperatures and associated high temperature gradients and larger lateral deflections of the wall as observed for gypsum plasterboard–lined walls. All three wall panels suffered severe board cracking on the ambient side (Figure 3.15) and failed by integrity criterion, giving only a 30 min FRL. The high mass loss caused board cracking in all three tests.

The use of noggings led to excessive board cracking as it restricted thermal expansion and bowing of the boards. When noggings were removed, joint opening and cracking due to excessive shrinkage and bowing of the boards caused panel failure (Figure 3.15). Fire rated sealants between boards and studs and in the joints could not withstand the excessive movement caused by board bowing and shrinkage. Strengthening the board joints with high temperature mortar also did not prevent joint cracking. Fibre mesh appeared to have been removed along the recessed board edges, which also affected the board strength in fire. This study has shown that mass loss characteristics of boards, joint detailing and stud spacing are important factors affecting the FRLs of LSF walls.

Hanna et al.'s (2015) fire tests of load-bearing walls also showed similar observations. In their tests, ambient side cracks were seen at 40 and 50 min, and they were wide open allowing the passage of hot gases. To eliminate the problems with MgO board, Chen et al. (2013a) suggested the use of MgO board as the inner layer and gypsum plasterboard as the outer layer. These studies seem to indicate that MgO boards are not an effective solution in improving the fire resistance of LSF walls.

FIGURE 3.15 Fire tests of MgO board-lined LSF walls. (Reprinted from *Fire Saf. J.*, 90, Rusthi, M. et al., Fire tests of magnesium oxide board lined light gauge steel frame wall systems, 15–27, Copyright 2017, with permission from Elsevier.)

New types of fire protective boards are being regularly introduced into the building sector. It is essential that full-scale fire tests are conducted to demonstrate that they can provide thin-walled steel structures with adequate FRLs. Furthermore, to facilitate advanced analyses, it is important that manufacturers and suppliers provide temperature-dependent thermal properties of their products such as specific heat, thermal conductivity and mass loss.

3.2.7 Effects of Steel Sheathing on Fire Resistance

Dias et al. (2019a) investigated the effects of using 0.55 mm thick G300 steel sheathing together with two layers of 16 mm gypsum boards on the FRLs of LSF walls made of 1.15 mm G500 steel web-stiffened studs. Three standard fire tests (Table 3.1) were conducted under a load ratio of 0.4, with no steel sheathing (Test No. 1), internal steel sheathing (Test No. 2) and external steel sheathing (Test No. 3).

The web-stiffened stud was designed to eliminate local, distortional and flexural-torsional buckling when used as 3 m wall studs with rigid sheathing on both sides (Dias et al. 2018). Hence the dominant failure mode was major axis flexural buckling of the studs. Small-scale non-load-bearing fire tests showed that when steel sheathing was used either internally or externally, the ambient side temperature was reduced and the insulation-based FRL was enhanced by about 16%. This was found to be due to retention, cyclic evaporation and recondensation of plasterboard moisture as a result of confinement provided by the steel sheathing (Dias et al. 2019b). In contrast, the addition of steel sheathing only provided marginal improvement to the FRL of load-bearing walls (Table 3.1) because both plasterboard and steel sheathing joints opened up and caused localised stud temperature rise, leading to structural failure, as seen in Figure 3.16.

Table 3.1 indicates that internally sheathed walls performed slightly better than externally sheathed walls (118 versus 111 min) because the internal sheathing joints are concealed and confined between the studs and plasterboards. However, slender steel–sheathing buckles prematurely between the screws when exposed to heat, exposing the plasterboard joints to high temperatures. In general, at the same load ratio (ratio of fire test load to ambient temperature capacity), significant FRL improvements are unlikely with internal or external steel sheathing.

TABLE 3.1 LSF wall configurations with steel sheathing and test results

TEST NO.	WALL CONFIGURATION	SHEATHING ARRANGEMENT		AMBIENT TEMPERATURE CAPACITY (KN)	FIRE TEST LOAD (KN)	FRL (MIN)	HF_{MAX} (°C)
		PLASTERBOARD	STEEL SHEET				
1		Double	–	118	47.2	110	417
2		Double	Internal	130	52.0	118	452
3		Double	External	130	52.0	111	414

Source: Reprinted from Thin-Walled Struct., 137, Dias, Y. et al., Full-scale fire tests of steel- and plasterboard-sheathed web-stiffened stud walls, 81–93, Copyright 2019, with permission from Elsevier.

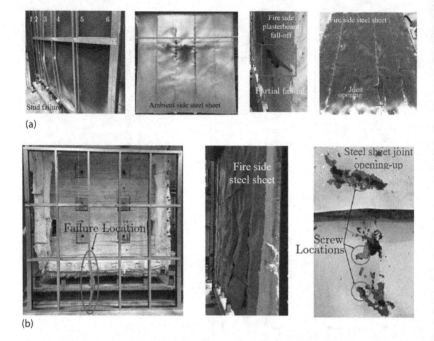

FIGURE 3.16 Failures of steel-sheathed LSF walls. (a) LSF wall with internal steel sheathing, (b) LSF wall with external steel sheathing. (Reprinted from *Thin-Walled Struct.*, 137, Dias, Y. et al., Full-scale fire tests of steel and plasterboard sheathed web-stiffened stud walls, 81–93, Copyright 2019, with permission from Elsevier.)

3.3 FIRE RESISTANCE TESTS OF FLOORS

3.3.1 General Behaviour of LSF Floors in Fire

Sultan et al. (1998), Alfawakhiri and Sultan (2001) and Alfawakhiri (2001) conducted five standard fire tests of 4.85 × 3.95 m floors lined with 15.9 mm thick plywood (in one test a layer of 51 mm thick concrete was added to the plywood) and one or two layers of 12.7 mm thick gypsum plasterboard under an applied loading in the range of 1.8–2 kN/m². Four of them had glass or rock fibre insulation in the cavity. Their tests showed that the number, thickness and type of fire protective boards and cavity insulation significantly influenced the FRLs of LSF floors whereas joist spacing (406 vs 610 mm) or type of sub-floor did not influence the FRL. As observed for LSF walls and for the same

FIGURE 3.17 Failures of LSF floor panels. (From Alfawakhiri, F. and Sultan, M.A., Loadbearing capacity of cold-formed steel joists subjected to severe heating, *Proceedings of the 9th International Conference Proceedings, Interflam 2001*, Edinburgh, UK, Vol. 1, pp. 431–442, 2001. With permission from National Research Council of Canada.)

reasons, the FRL of load-bearing floors was significantly reduced by the use of cavity insulation. Integrity and insulation failures were not observed, and the floor panel collapsed due to section failure of the joists (Figure 3.17).

Sakumoto et al. (2003) conducted six fire tests of 4.26 × 2.95 m LSF floors made of back-to-back lipped channel joists lined with composite boards including plywood on the top and one or two layers of gypsum plasterboards of thicknesses 9.5, 12.5 and 15 mm. Their tests also showed that the FRL increased significantly with increasing thickness and number of boards, but cavity insulation did not reduce the FRL as expected. Plasterboard fall-off was identified as the key factor influencing the FRL.

Zhao et al. (2005) conducted full-scale fire tests of two floor systems of 5.5 × 2.99 m made of 250 × 2.5/2.0 mm lipped channel joists at 600 mm spacing and lined with two layers of gypsum board and cavity insulation. With fire on one side, non-uniform temperature distribution was developed, which caused thermal bowing deformations in addition to deflections caused by the applied load. With increasing temperatures on the fireside and plasterboard fall-off, the joists failed in bending due to reduced mechanical properties in fire. Localised web crippling failures were also observed at the joist to track supports.

Baleshan and Mahendran (2016) conducted three fire tests of LSF floors made of 180 × 1.15 mm lipped channel joists with two layers of 16 mm plasterboard on the fireside under a load ratio of 0.4. Rock fibre insulation was used in the cavity in the second test while it was used between the two plasterboards as external insulation in the third test, following the recommendation of Kolarkar and Mahendran (2008, 2012). In all three tests, the joists moved towards the fireside and failed locally by web crippling near the supports. The use of cavity insulation reduced the failure time from 107 to 99 min due to higher hot flange temperatures and larger lateral deflections caused by greater temperature gradients across the floor depth. On the other hand, the use of external insulation increased the failure time to 139 min. These observations are similar to those observed for LSF walls as discussed in Sections 3.2.2 and 3.2.3.

From the aforementioned standard fire tests of LSF floors, the following useful observations are made:

- LSF floors are generally made of single or back-to-back thin-walled steel-lipped channel joists of thicknesses in the range of 1.15–2.5 mm and made of steels with yield strengths up to 550 MPa.
- Protecting them with one or two layers of fire rated gypsum plasterboard with thicknesses in the range of 9.5–16 mm has been successful in providing the required FRLs. Significant improvement to FRLs can be achieved by using thicker plasterboards and/or more than one layer.
- Plasterboard fall-off can significantly reduce the FRLs of LSF floors. This is more critical for LSF floors than for LSF walls.
- Cavity insulation is commonly used, but it will reduce the FRL of load-bearing LSF floors.
- Joist failure is governed mostly by the average joist temperature. It is usually based on a combination of local buckling of the compression cold flanges and yielding of the hot flanges.
- Increasing load ratios leads to lower FRLs as for LSF walls.

3.3.2 Effects of New Joist Sections on Fire Resistance

Jatheeshan and Mahendran (2016b) investigated the possibility of increasing the fire resistance of LSF floors by using a new welded hollow flange channel (HFC) section of dimensions 200 × 45 × 15 × 1.6 mm as joists. When the results of full-scale fire tests of floor panels lined with two 16 mm plasterboards with and without cavity insulation were compared with those from Baleshan and Mahendran (2016), significant improvements to fire resistance times were noted. However, as with LSF walls, the improvement was a result of the enhanced elevated temperature mechanical properties of HFC due to

FIGURE 3.18   Failure modes of LSF floors made of HFC joists.

the simultaneous cold-forming and dual electric resistance welding fabrication process used and the possible improved connectivity of joist to plasterboards through both flanges. Cavity insulation adversely affected the failure times as expected. Plasterboard joint and fall-off (Figure 3.18) affected the temperature development in the joists and influenced the failure times significantly. The failure of HFC joist was due to local buckling of the compression cold flanges and significant yielding of the hot flanges with large lateral deflections as seen in Figure 3.18.

3.4 OTHER FIRE RESISTANCE TESTS ON THIN-WALLED STEEL STRUCTURES

Thin-walled steel structural elements and assemblies are normally protected by fire rated boards and are thus subject to non-uniform temperature distributions when exposed to fire conditions as discussed in the previous sections. However, in some applications such as columns surrounded by fire from all sides, they can experience nearly uniform temperature distributions. Their behaviour under uniform temperature condition also forms the basis of developing understanding for the more complex condition of non-uniform temperature distribution. For these reasons, many experimental studies have been undertaken on unprotected thin-walled steel columns and beams under uniform elevated temperature conditions. This section presents brief details of selected experimental studies and their results.

Feng et al. (2003a, 2003b) experimentally and numerically investigated both local and distortional buckling capacities of short lipped and unlipped channel columns at uniform elevated temperatures including the effects of web holes. Gunalan et al. (2015) investigated local buckling behaviour of short thin-walled steel columns under uniform elevated temperatures up to 700°C. Both unlipped and lipped channel sections made of low- and high-strength steels were considered. Similarly, Ranawaka and Mahendran (2009a) investigated the distortional buckling behaviour of two types of lipped channel columns (with and without additional lips) made of low- and high-strength steels exposed to uniform elevated temperatures up to 800°C. Figure 3.19 shows the typical test set-up used in their tests. The test results clearly showed reductions in the local and distortional buckling capacities with increasing temperatures.

These results also indicated that the current ambient temperature design equations based on the effective width method or the direct strength method (DSM) in AISI S100 (AISI 2016), Eurocode 3 Part 1.3 (CEN 2006a) and AS/NZS 4600 (SA 2018) can be used to conservatively predict the local buckling capacities by using appropriately reduced mechanical properties

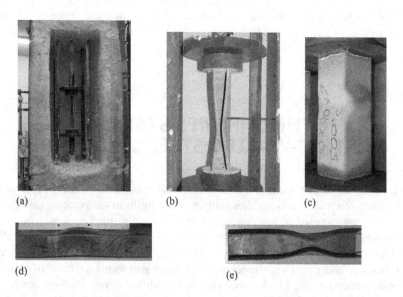

(a) (b) (c)

(d) (e)

FIGURE 3.19 Uniform elevated temperature tests and failure modes of short thin-walled steel columns: (a) electrical furnace, (b) distortional buckling, (c) local buckling of SHS, (d) local web buckling of lipped channel and (e) local flange buckling of unlipped channel. (Reprinted from *J. Constr. Steel Res.*, 65, Ranawaka, T. and Mahendran, M., Distortional buckling tests of cold-formed steel compression members at elevated temperatures, 249–259, Copyright 2009, with permission from Elsevier.)

in all the calculations. The DSM-based equations for distortional buckling at ambient temperature were also able to predict the elevated temperature capacity reasonably well if appropriately reduced mechanical properties were used.

Bandula Heva and Mahendran (2013) and Gunalan et al. (2014) investigated the flexural-torsional buckling behaviour of long thin-walled lipped channel columns under uniform elevated temperatures up to 700°C using an electric furnace as shown in Figure 3.20. In general, ambient temperature design rules from AS/NZS 4600 (SA 2018), AISI S100 (AISI 2016) and Eurocode 3 Part 1.3 (CEN 2006a), with a suitable modification of the buckling curves, accurately predicted the reduced elevated temperature capacities of pin-ended slender lipped channel columns, provided appropriately reduced mechanical properties are used. For fix-ended columns, AISI S100 (AISI 2016) and AS/NZS 4600 (SA 2018) design equations were modified by Gunalan et al. (2013) to include the beneficial effect of warping restraint and were then used for elevated temperature conditions successfully.

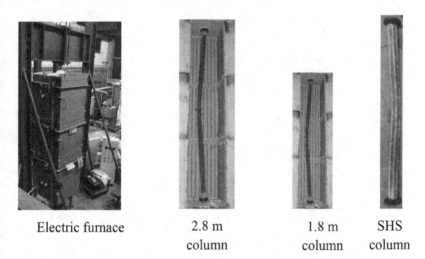

Electric furnace   2.8 m column   1.8 m column   SHS column

FIGURE 3.20 Uniform elevated temperature tests of long thin-walled steel columns. (Reprinted from *Eng. Struct.*, 79, Gunalan, S. et al., Flexural-torsional buckling behaviour and design of cold-formed steel compression members at elevated temperatures, 149–168, Copyright 2014, with permission from Elsevier.)

3.5 FINAL COMMENTS

This chapter has provided the details of a large number of standard fire tests on LSF walls and floors and uniform elevated temperature tests on thin-walled steel sections. The test results were used to explain their behaviour and how key parameters influence their fire resistance. An important focus of this chapter was to identify methods of enhancing the fire resistance of load-bearing and non-load-bearing LSF structures. The next chapter will describe how to numerically evaluate the fire resistance of LSF structures.

Numerical Modelling of Fire Resistance of Thin-Walled Steel Structures

4

In order to evaluate the fire resistance of thin-walled structures at elevated temperatures, it is necessary to understand the fire behaviour and to perform heat transfer analysis to obtain the temperatures of the fire exposed structures. It is also necessary in elevated temperature structural analysis to include reduced mechanical properties and additional thermal effects on structural performance such as thermal expansion and thermal bowing. This chapter summarises the key features of numerically performing fire resistance evaluation and provides some simplified solutions.

4.1 FIRE BEHAVIOUR

The behaviour of fires in compartments is complex. For structural fire engineering applications, simplifying assumptions have to be made. These simplifying assumptions lead to gas/fire temperature–time relationships to be used in thermal boundary conditions for heat transfer analysis.

Thin-walled steel structures are typically used in buildings to enclose relatively small spaces. Therefore, it is acceptable to treat flashover enclosure

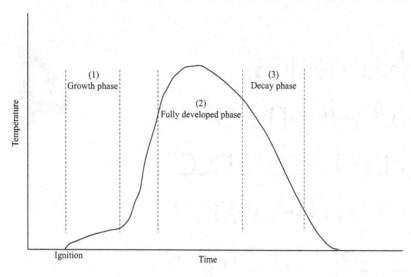

FIGURE 4.1   Enclosure post-flashover fire temperature–time relationship.

fires as uniformly distributed. If uninterrupted, such an enclosure fire goes through three phases as illustrated in Figure 4.1.

1. Growth phase until flashover: Immediately after ignition the fire begins to grow. In this phase, the fire is localised in the enclosure with cooler air near the floor of the compartment and smoke and hotter air accumulating beneath the ceiling. This lasts until the hot smoke and fire radiates sufficient heat to pyrolysis all other combustible materials to cause ignition within a very short period of time, this event being termed flashover.

2. Fully developed (post-flashover) phase: After flashover, the heat release rate of the fire reaches the maximum, either limited by the supply of oxygen through the openings or available fuel in the compartment. This phase is the most critical stage to structural stability because very high temperatures are involved.

3. Decay or cooling phase: When nearly all combustible materials are consumed, the fire enters the decay phase with decreasing temperature. This phase may still be important for the structure because cooling may induce tensile forces in some structural members if they are restrained. However, this book focuses on isolated structural members, as is commonly practised. The effects of structural restraints are not considered.

For assessment of the behaviour of structures in fire, the growth phase is ignored as their temperatures are usually low.

4.1.1 Nominal Fires

Precise quantification of enclosure fire behaviour remains a scientific challenge. For practical fire resistance design, simplified analytical models are available to calculate the fire temperature–time relationships. The most common analytical models are based on nominal fire curves and parametric fire curves.

The nominal fire curves define a single temperature–time relationship independent of realistic fire conditions. They were developed for grading fire resistance of construction elements and are adopted in design standards worldwide with some minor differences. However, the nominal fire curves have a number of shortcomings, including:

1. They do not represent real fire behaviour.
2. They do not always give the most severe fire condition.
3. They do not have a cooling phase.

Table 4.1 lists the most commonly used nominal fire conditions and their parameters.

The nominal fire temperature–time curves are:

- Standard fire curve (ISO 1999, CEN 2002, SA 2005):

$$T_f = 20 + 345 \log 10 \left(8t + 1 \right) \tag{4.1}$$

- External fire curve (CEN 2002):

$$T_f = 660 \left(1 - 0.687 \, e^{-0.32t} - 0.313 \, e^{-3.8t} \right) + 20 \tag{4.2}$$

- Hydrocarbon curve (CEN 2002):

$$T_f = 1080 \left(1 - 0.325 \, e^{-0.167t} - 0.675 \, e^{-2.5t} \right) + 20 \tag{4.3}$$

where:
   T_f is the gas/fire temperature in °C
   t is time in min

Figure 4.2 compares the different nominal fire temperature–time curves.

TABLE 4.1   Nominal fire curves and their applications

FIRE TYPE	APPLICATION	COEFFICIENT OF HEAT TRANSFER BY CONVECTION (W/m²K)
Standard fire	Fully developed cellulosic fire inside fire enclosure	25
External fire	Fully developed cellulosic fire outside fire enclosure	25
Hydrocarbon fire	Fully developed hydrocarbon fire inside enclosure	50

Source: European Committee for Standardization (CEN), Eurocode 1: Actions on structures – Part 1–2: General actions – Actions on structures exposed to fire, Brussels, Belgium, 2002.

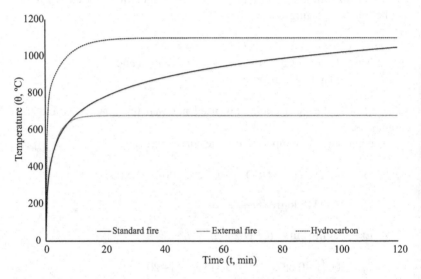

FIGURE 4.2   Comparison of nominal fire temperature–time curves.

4.1.2 Parametric Fires

A more realistic representation of fires in enclosures is the parametric fire curve according to EN 1991-1-2 (CEN 2002). The parametric fire temperature-time curve is a function of fire load, ventilation condition and compartment characteristics. Figure 4.3 shows an example of parametric fire curve, consisting of a heating phase until the maximum temperature followed by a

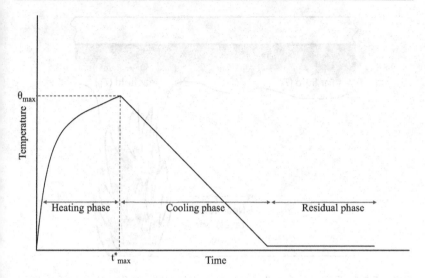

FIGURE 4.3   An example of parametric fire temperature–time curve.

linearly decreasing cooling phase until the ambient temperature. Complete details of how to quantify a parametric fire curve are given in EN 1991-1-2 (CEN 2002).

4.1.3 Travelling Fires

Recently, a considerable amount of research studies have been devoted to travelling fires, as they have been observed in large compartment fire tests and fire accidents. A travelling fire consists of a near field (flame) and a far field (smoke), with different temperatures (Stern-Gottfried and Rein 2012), as illustrated in Figure 4.4. The size of the near field can change and can be considered a design variable. Due to non-uniformity of heating across a compartment, the behaviour and failure mechanism of structural elements can be very complex. For applications of thin-walled steel structures in relatively small enclosures where the fire temperature can be considered uniform, it is not necessary to consider travelling fires.

4.1.4 Computer Programs

Should it become necessary, a number of computer programs are available for modelling compartment fires. However, using computer programs to model

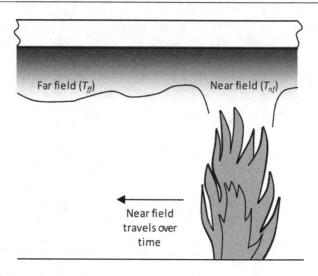

FIGURE 4.4   Travelling fire. (From Stern-Gottfried, J. and Rein, G., *Fire Saf. J.*, 54, 96–112, 2012. With permission.)

enclosure fire behaviour requires special expertise and detailed fire dynamics knowledge. They are rarely used by structural engineers.

Enclosure fires may be modelled using zone models, where the enclosure is divided into a very small (typically 2) number of zones, each having uniform properties or computational fluid dynamics (CFD) models. Popular zone and CFD models for modelling enclosure fires include the following (Wang et al. 2012):

- COMPF2 (Babrauskas 1979): 'C0MPF2 is a computer program for calculating the characteristics of a post-flashover fire in a single building compartment (single zone), based on fire-induced ventilation through a single door or window. It is intended both for performing design calculations and for the analysis of experimental burn data.'
- OZone V2 (Cadorin et al. 2003): The OZone V2 software calculates the evolution of gas temperature in a compartment under fire and evaluates the fire resistance of single steel elements. It is based on one- or two-zone models, with an automatic shift from two to single zone model under certain circumstances.
- SFIRE-4 (Magnusson and Thelandersson 1970): The model takes into consideration the amount of water discharged by sprinklers, etc. The bounding structures of the enclosure can be composed of

up to three different wall constructions or systems. This program is based on single zone model.

- CCFM (Cooper and Forney 1990): This program was developed for multi-room fire behaviour based on two-zone models. CCFM refers to Consolidated Compartment Fire Model (CCFM).
- CFAST (Peacock et al. 2017): CFAST is a two-zone fire model used to calculate the evolving distributions of smoke, fire gases and temperature throughout compartments of a building during fire.
- Fire Dynamics Simulator (FDS) and Smokeview (SMV) (McGrattan et al. 2018): FDS is a CFD model of fire-driven fluid flow. The software solves numerically a form of the Navier-Stokes equations appropriate to low-speed, thermally driven flow, with an emphasis on smoke and heat transport from fires. FDS is commonly used by fire protection engineers.

4.2 HEAT TRANSFER MODELLING

The theory of heat transfer is well established. The main issues are to acquire reliable thermal material properties and to develop simplified models for practical design. Chapter 5 presents relevant material properties. This section describes the fundamental heat transfer equations and their implementation in a simplified heat transfer model for thin-walled steel structural panels.

4.2.1 Basics of Heat Transfer

The basic equation for conductive heat transfer equation is:

$$\dot{Q}_{cond} = -kA\frac{dT}{dx} \tag{4.4}$$

where, referring to Figure 4.5:

A is the area across which heat is transferred [m^2]
k is the thermal conductivity of the material [W/m K]
T is temperature [K]
x is the distance normal to the area A [m]
dT/dx is temperature gradient

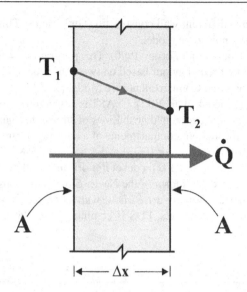

FIGURE 4.5 Heat transfer within a member.

For transient heat transfer in three-dimensions in the absence of internal generation of heat, the conductive heat transfer equation is (Cengel and Ghajar 2014):

$$\frac{\partial}{\partial x}\left(k\frac{\partial T}{\partial x}\right)+\frac{\partial}{\partial y}\left(k\frac{\partial T}{\partial y}\right)+\frac{\partial}{\partial z}\left(k\frac{\partial T}{\partial z}\right)=\rho C_p\frac{\partial T}{\partial t} \qquad (4.5)$$

where:
  t is time [sec]
  ρ is the density of the material [kg/m³]
  C_p is the specific heat of the material [J/kg K]

4.2.2 Thermal Boundary Conditions for Heat Transfer Modelling

In order to solve the general conductive heat transfer Eq. (4.5), thermal boundary conditions are required. For evaluating structure fire resistance, the thermal boundary conditions consist of convective heat transfer and radiant heat transfer boundary conditions.

The convective heat transfer boundary condition is expressed as (Cengel and Ghajar 2014):

$$\dot{Q}_{conv} = h_{conv} A_s \left(T_f - T_s\right) \tag{4.6}$$

where:

$\dot{Q}_{conv}$ is the rate of convective heat transfer (W)

h_{conv} is the convective heat transfer coefficient in W/m²°C

A_s is the surface area through which convective heat transfer takes place

T_s is the surface temperature [°C]

T_f is the temperature of the gas sufficiently far from the surface [°C] (fire temperature)

The radiation heat transfer boundary condition is (CEN 2002):

$$\dot{Q}_{rad} = \varepsilon\sigma A_s \left(\left(T_f + 273.15\right)^4 - \left(T_s + 273.15\right)^4\right) \tag{4.7}$$

where:

$\dot{Q}_{rad}$ is the rate of radiant heat transfer (W)

ε is the emissivity of the surface. The property emissivity, whose value is in the range between 0 and 1, is a measure of how closely a surface approximates a blackbody surface, which has an emissivity value of 1 ($\varepsilon = 1$)

σ is the Stefan–Boltzmann constant ($\sigma = 5.67 \times 10^{-8} \text{W/m}^2 \cdot \text{K}^4$)

A_s is the surface area through which radiant heat transfer takes place

T_s is the surface temperature

T_f is the surrounding air (fire) temperature

Figure 4.6 illustrates different modes of heat transfer for a thin-walled steel structural panel with fire exposure from one side (Shahbazian and Wang 2014b).

4.2.3 Computer Programs

In general, there is no analytical solution to three-dimensional transient heat transfer problems. Therefore, numerical simulation methods are used. Any general finite element package, such as ABAQUS and ANSYS, will be able to

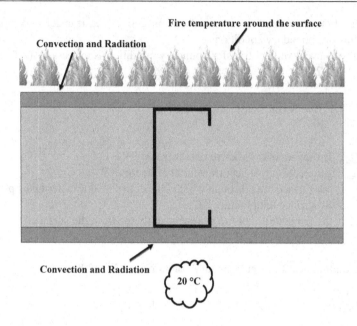

FIGURE 4.6   Thermal boundary conditions.

perform heat transfer simulations. There are also a few specialist heat transfer programs that have been specifically developed for performing heat transfer calculations for structures under fire conditions, although some were developed a long time ago and are not maintained. They include FIRES-T3 (Iding et al. 1977), TASEF (Gerlich et al. 1996), WALL2D (Takeda and Mehaffey 1999) and SAFIR (Franssen and Gerney 2017).

4.2.4 A Simplified Heat Transfer Model for Thin-Walled Steel Structural Panels

Whilst the above-mentioned general finite element packages and specialist heat transfer programs are powerful and flexible modelling tools, their applications require specialist expertise. To facilitate rapid and accurate calculations of heat transfer in thin-walled steel structural panels subjected to fire from one side, Shahbazian and Wang (2013) developed a simplified procedure that is suitable for implementation in Excel or other programs (An example is available on MATLAB File Exchange (2013)). This procedure is described next.

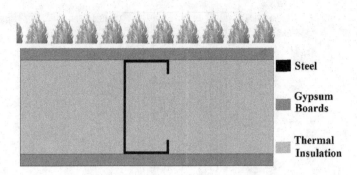

FIGURE 4.7 Schematic view of a thin-walled steel structural panel.

Figure 4.7 shows the problem under consideration. In the simplified procedure, the two-dimensional heat transfer problem is converted to a one-dimensional
problem by using equivalent thermal resistances for the combined steel section
and cavity insulation. This is possible because the research studies by Feng et al.
(2003d) and Shahbazian and Wang (2013) have concluded that the steel section
temperature distribution can be assumed to be linear through the depth of
the panel and the average flange temperatures can be used (see Figure 3.5).

Figure 4.8 shows a schematic view of the simplified model including thermal boundary conditions and the thermal network for heat transfer and calculating heat capacity. For this scheme, the gypsum board on each side is divided into
five layers (1–5, 11–15) and the steel section/insulation is also divided into five
layers (6–10). In the simplified method, heat transfer in the wall panel containing
a steel cross-section is assumed to be one dimensional through the depth of the
panel. For one layer, the following heat balance equation may be written:

$$
\begin{pmatrix} \text{heat} \\ \text{entering} \\ \text{the layer} \end{pmatrix} - \begin{pmatrix} \text{heat} \\ \text{leaving} \\ \text{the layer} \end{pmatrix} = \begin{pmatrix} \text{heat in the layer} \\ \text{to change the layer} \\ \text{temperature} \end{pmatrix} \tag{4.8}
$$

Over a small time interval, the heat transfer between any two layers can be
generally written as:

$$
\dot{Q} = \frac{\Delta T}{\Sigma R} \tag{4.9}
$$

where:

ΔT is the temperature difference between these two layers
ΣR is the total thermal resistance in the heat transfer path

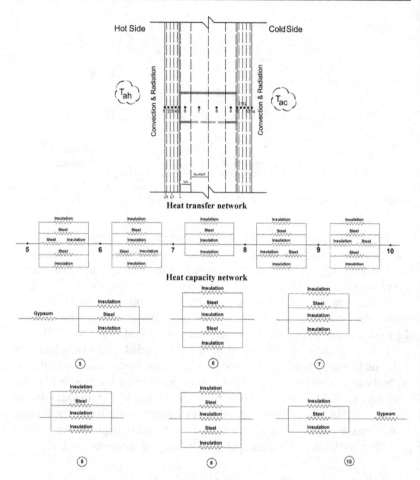

FIGURE 4.8 Schematic view of the simplified heat transfer model.

The amount of heat required to change the temperature of a layer is:

$$\dot{Q}_{cap} = \sum R_{cap} \frac{dT}{dt} \tag{4.10}$$

where:

t is time

$\sum R_{cap}$ is the total heat capacitance (mass times specific heat) of the layer

In this simplified method, the thermal resistance is calculated using the weighted average of the materials within the heat transfer path. Details are as follows:

- For heat transfer between the fire and the slice of gypsum board on the fire exposed side (point 1), the thermal resistance is the total of the thermal boundary layer and ½ of the gypsum slice.
- For heat conduction between two adjacent gypsum slices (points 2–3, 3–4, 4–5, 10–11, 11–12 and 12–13), the thermal resistance is the total of the two halves of each slice.
- For heat transfer between the slice of gypsum board on the air side with the air layer, the thermal resistance is the total of the air layer and ½ of the gypsum slice (point 14).
- For heat transfer between a gypsum slice and a steel slice or between two adjacent steel slices, calculation of the total thermal resistance should include different materials that form the slices. For each slice, the thermal resistances are in parallel.
- To calculate the heat capacitance of each slice, all the materials within that slice should be included.

To calculate the thermal resistance and heat capacitance, it is necessary to determine an effective width of the panel that should be included in the calculations (for one-dimensional assumption). The effective width of the panel (W_e) used for calculating the weighted average of thermal resistance is approximately:

$$W_e = 45 + 0.85 b_f \ (\text{in mm}) \tag{4.11}$$

where b_f is the flange width of the thin-walled steel section.

4.3 MODELLING BEHAVIOUR OF THIN-WALLED STEEL STRUCTURES AT ELEVATED TEMPERATURES

The calculation of stresses and failure modes in thin-walled steel structural members is a complex procedure even at ambient temperature. Only very brief guidance is provided in this section.

When a thin-walled structure is subjected to compressive stress, different modes of buckling may occur. Elastic buckling is characterised by a sudden out-of-plane deflection and it occurs well below the yield stress of the material. In modelling of thin-walled structures, the elastic buckling modes, which can be obtained by performing eigenvalue analysis using finite element software such as ABAQUS or finite strip programs such as CUFSM, are used to specify initial geometric imperfections for nonlinear analysis. In order to apply initial geometric imperfection, eigenvalue analysis should be performed first in order to find the lowest buckling modes (local, distortional or global) and any inter-action between different modes. Then based on sensitivity study and numerical validation, imperfection scale/magnitude should be defined in nonlinear numerical analysis. However, in general, the local initial imperfection scale, global initial imperfection scale and distortional initial imperfection scale may be assumed to be equal to the thickness of the plate (Feng 2003), L/500 (where L is the column length) (Kaitila 2002) and half of the thickness of the plate (Shahbazian and Wang 2012), respectively.

When conducting numerical simulation of structural behaviour in fire, steady state or transient state analysis may be used. In steady state analysis, the temperatures of structure are increased to the pre-determined level and this is then followed by increasing the mechanical load. In transient state analysis, the mechanical load is applied first and maintained, followed by increasing the temperatures of structures. The steady state analysis is easier to implement. However, for modelling structural behaviour in fire, the transient state analysis should be used because this mimics the situation of structural behaviour in fire.

4.4 SUMMARY

The distribution of temperatures in thin-walled steel sections as part of a panel exposed to fire on one side can be complex. However, for evaluations of the load carrying capacity of LSF structures, the temperature profile can be simplified. Furthermore, the two-dimensional heat transfer problem can be approximately reduced to a simple one-dimensional problem by using the weighted average of thermal properties of the different constituent parts of the thin-walled steel structural panel.

Elevated Temperature Properties of Materials

5

5.1 INTRODUCTION

In fire engineering calculations of structural resistance in fire, it is necessary to have accurate and reliable information on thermal and mechanical properties of materials of the structure. Compared to hot-rolled steel structures, there are more issues for the material properties of cold-formed thin-walled steel structures. Cold-formed steel structures are made by cold working of thin-walled cold-rolled steel strips into structural shapes. This introduces strain hardening around corners, and the amount of strain hardening may be affected by the thickness of the structure. Lightweight board and insulation materials, such as gypsum plasterboard and mineral fibres, are used in thin-walled steel construction systems. The thermal properties, in particular thermal conductivity, of these materials are not constant. Since temperatures attained in steel sections of thin-walled construction are critically dependent on insulation properties, thermal properties of insulation materials and gypsum plasterboards should be provided. This chapter will review available data from major sources of relevant information in literature and make recommendations.

5.2 MECHANICAL PROPERTIES OF COLD-FORMED STEELS AT ELEVATED TEMPERATURES

Cold-forming of steel does not change its composition, therefore, the thermal properties of cold-formed steel are the same as hot-rolled steel. It is also expected that the elastic modulus of cold-formed steel is the same as hot-rolled steel. Results of mechanical property tests by various investigators have revealed some differences in elevated temperature modulus of elasticity values, but because the modulus of elasticity value is a sensitive quantity for measurement, these differences can usually be attributed to differences in test setup and measurement. Therefore, this section will focus on changes in effective yield strength of cold-formed steel at elevated temperatures.

Chen and Young (2006) reported experimental results of mechanical properties of cold-formed steel. They compared elevated temperature mechanical properties of steel coupons taken from the corner and flat parts of steel sections. The steel was supplied by BlueScope Lysaght (Singapore) and the steel grade was G500 (minimum yield strength of 500 MPa). The elevated temperature tests were carried out under steady state condition. They found that the corner and flat coupons had very similar results for elevated temperature mechanical property reduction factors.

Chen and Young (2007) provided further experimental results for BlueScope Lysaght steel grades G550 (1.0 mm) and G450 (1.9 mm) under both steady state and transient state conditions. Their results seem to be very different from the results of others, with much higher modulus of elasticity but much lower yield stress at temperatures above about 500°C. Their results are also quite different from their own test results reported in Chen and Young (2006) for a slightly different grade of steel (G500). These make it difficult to use their results to identify any discernible trend.

Ye and Chen (2013) carried out similar tests as Chen and Young (2006) on Q345 steel, also supplied by BlueScope Lysaght (Shanghai), and came to the same conclusion with regard to corner and flat plate steels. They compared the test results under steady and transient state test conditions and found that transient state testing gave similar results at temperatures not exceeding 400°C and higher yield strengths at higher temperatures. Similar results have been presented by Chen and Ye (2012) for G550 steel. However, even their transient test results of yield strength reduction factors of steel were lower than those recommended in EN 1993-1-2 (CEN 2005). The EN 1993-1-2 reduction factors for cold-formed steel at elevated temperatures are the same as the 0.2% proof stress of hot-rolled steel. They are very close to those of Outinen (1999).

Ranawaka and Mahendran (2009b) provided experimental results for three thicknesses (0.6, 0.8 and 0.95 mm) of two grades of cold-formed steel (G550 and G250) under steady state condition. Their results indicate that the effect of thickness is minor. However, the two grades of steel have different yield strength reduction factors, with the values for G550 steel much greater than those of G250 steel at temperatures lower than 550°C, but much lower at temperatures above 550°C.

Kankanamge and Mahendran (2011) performed further mechanical property tests of G250 and G450 steels for two different thicknesses (1.50 mm/1.55 mm, 1.90 mm/1.95 mm). They reached similar conclusions as Ranawaka and Mahendran (2009b) described above and proposed improved predictive equations by combining with Ranawaka and Mahendran's (2009b) test results. Both research studies concluded that the thickness of steel in the range of 0.55–1.95 mm has minor influence on the mechanical properties of cold-formed steel at elevated temperatures.

A review paper by Craveiro et al. (2016) confirmed that the values of modulus of elasticity of cold-formed steel obtained by different researchers for different cold-formed steels, except for those of Chen and Young (2007), were reasonably consistent. The recommendations in EN 1993-1-2 (CEN 2005) tend to be at the upper bound of various test results at temperatures not exceeding 600°C but at the lower bound of test results at temperatures higher than 600°C.

On the other hand, there are large differences in yield strength results from different researchers across the whole elevated temperature range of interest to fire resistance. The recommended yield strength reduction factors in EN 1993-1-2 (CEN 2005) for cold-formed steel are close to the average of results by various researchers.

To summarise, test results by different researchers show large differences in yield strength reduction factors of cold-formed steels at elevated temperatures. However, these differences are not caused by steel thickness or location (corner/flat) of steel in the cross-section. The results obtained under the steady and transient state conditions can be different for the same steel, with transient state results seeming to be higher than steady state test results. The most influential factor seems to be steel grade/manufacturer. This makes it difficult to propose a universally acceptable set of yield strength reduction factor – temperature relationship for different cold-formed steels. Ideally, specific elevated temperature mechanical property tests should be carried out to obtain data for the specific grade and type of steel. For example, the relationship in AS/NZS 4600 (SA 2018) should only be applied to Australian/NZ cold-formed steels. Nevertheless, the recommended values in EN 1993-1-2 represent approximately the average of results from different researchers for different grades and types of steel. Therefore, where specific information is not available, this book recommends using the EN 1993-1-2 values.

The recommended reduction factors for modulus of elasticity and effective yield strength of cold-formed steel at elevated temperatures are listed in Table 5.1.

TABLE 5.1 Reduction factors for yield strength and modulus of elasticity of cold-formed steel at elevated temperatures

TEMPERATURE (°C)	20	100	200	300	400	500	600	700	800	900	1000	1100	1200
Yield strength	1	1	0.89	0.78	0.65	0.53	0.30	0.13	0.07	0.05	0.03	0.02	0.0
Modulus of elasticity	1	1	0.9	0.8	0.7	0.6	0.31	0.13	0.09	Linear interpolation			0.0

Source: European Committee for Standardization (CEN), EN 1993-1-2, Eurocode 3: Design of steel structures – Part 1.2: General rules – structural fire design, Brussels, Belgium, 2005.

5.3 THERMAL PROPERTIES OF FIRE PROTECTION MATERIALS AT ELEVATED TEMPERATURES

5.3.1 General

In cold-formed thin-walled steel panel construction for walls and floors, lightweight fire protection and interior/exterior insulation materials are used to provide sufficient fire resistance and thermal/sound insulation. Accurate data of thermal properties of these materials are critical for safe and reliable calculation of fire resistance of these systems. In fact, since the mechanical properties of cold-formed steel decrease sharply at elevated temperatures in the region of interest, small errors in the calculation of steel temperature results can result in very large errors of calculating structural resistance. For example, when the steel temperature is increased from 500°C to 600°C (20% increase), according to the EN 1993-1-2 recommendation of reduction factors of steel yield strength and modulus of elasticity in Table 5.1, the yield strength and modulus of elasticity of steel decrease by 43% and 48%, respectively.

The thermal properties of a material include density (ρ), specific heat (C_p) and thermal conductivity (k). The product of density and specific heat (ρC_p) is thermal capacitance of the material, measuring the amount of heat (energy) required to raise the temperature of per unit volume of material by 1 degree. Since fire protection and insulation materials used in cold-formed thin-walled steel structures are lightweight, the accuracy of calculating steel temperatures is insensitive to even large changes in thermal capacitance of fire protection/insulation materials. On the other hand, fire protection and insulation materials have low thermal conductivities, therefore, temperatures attained in the steel structure vary almost linearly with variations in thermal conductivity of protection/insulation materials.

The thermal conductivities of fire protection and insulation materials at ambient temperature can be expected to be provided by the material manufacturers because they have to provide such information for evaluating the thermal insulation performance of the panels. However, the ambient temperature values should not be used in fire engineering calculations. This is because the thermal conductivity of lightweight materials increases with temperature.

As mentioned above, small errors of underestimation of steel temperature can lead to large errors of overestimation of steel mechanical properties, thus resulting in grossly unsafe design.

In lightweight materials, air voids exist. Therefore, void radiation at high temperatures accelerates heat transfer through the material. Since the power of radiation is related to T^4, the rate of heat transfer (thermal conductivity) through the material is related to T^3. Therefore, the thermal conductivity–temperature relationship of a lightweight material may be expressed as follows:

$$k = k_0 + Const \cdot T^3 \tag{5.1}$$

where:

k and k_0 are thermal conductivity of the material at temperature T and ambient temperature, respectively

T is the material temperature

Wang et al. (2012) carried out an assessment of publicly available information of thermal properties, in particular, thermal conductivities of some fire protection materials at elevated temperatures. They recommended the thermal property models in Table 5.2. It can be seen in Table 5.2 that the thermal conductivities of calcium silicate and vermiculite depend on their densities. This is because the higher the density, the lower the air void, thus the lower the void radiation. In fact, when their densities reach their complete solid densities of about 2540 kg/m³ and 1000 kg/m³, respectively, heat transfer due to void radiation stops because there is no void in the material. On the other hand, since the ambient temperature thermal conductivity of air is negligible compared to that of the material, the thermal conductivity of the material at ambient temperature k_0 increases linearly with increasing density.

5.3.2 Gypsum

Gypsum plasterboards are commonly used in cold-formed thin-walled steel panels as fire protection. Due to the presence of a large amount of free and chemically bound water, their thermal properties are affected by water movement and evaporation.

The total amount of water is about 24% by weight, consisting of about 21% of chemically bound water and 3% of free water. Water evaporation occurs in two stages, the first stage starting at about 100°C, during which about 75% of the chemically bound water and free water are lost. The temperature range over which the second stage of water evaporation occurs is still subject to

TABLE 5.2 Thermal property models for some common types of insulation material

MATERIAL	DENSITY ρ (kg/m³)	BASE VALUE OF SPECIFIC HEAT (J/kg·K)	THERMAL CONDUCTIVITY (W/m·K)
Rock fibre	155–180	900	$k = 0.022 + 0.1475\left(\dfrac{T}{1000}\right)^3$
Mineral wool	165	840	$k = 0.03 + 0.2438\left(\dfrac{T}{1000}\right)^3$
Calcium silicate	Various	900	$k = k_0 + Const\left(\dfrac{T}{1000}\right)^3$ $k_0 = 0.23\dfrac{\rho}{1000}$ $Const = 0.08\dfrac{2540 - \rho}{2540}$
Vermiculite	Various	900	$k = k_0 + Const\left(\dfrac{T}{1000}\right)^3$ $k_0 = 0.27\dfrac{\rho}{1000}$ $Const = 0.18\dfrac{1000 - \rho}{1000}$

Source: With kind permission from Taylor & Francis: *Performance-Based Fire Engineering of Structures*, 2012, Wang, Y.C. et al.

debate, with some stating immediately following completion of the first stage at about 200°C and others saying as late as 600°C. Because the amount of water involved in the second stage is relatively low, it is acceptable to assume that water evaporation is complete in the temperature range between 100°C and 200°C.

Based on the above assumption of water evaporation, the thermal properties of gypsum plasterboard can be quantified as follows.

5.3.2.1 Density

The density of gypsum plasterboard is 100% of its ambient density until 100°C, then linearly changing to 76% (100%–24%) at 200°C and maintaining at this value at higher temperatures.

5.3.2.2 Specific heat

Dehydration of gypsum is an endothermic process which consumes heat (energy). Exact evaluation of the temperature field in gypsum plasterboard, and hence the steel structure, will require modelling of combined heat and mass transfer. An alternative and simplified approach is to deal with only the heat transfer process and using an equivalent (increased) specific heat to take into consideration the additional heat required in dehydration. The additional heat of dehydration of gypsum plaster consists of three parts: the amount of heat required to break the chemical bond to release the chemically bond water, heat required to drive water from inside gypsum to the surface and the latent heat of water for evaporation. The latent heat of water is 2260 MJ/(kg·C). If the total amount of water is 'e', then the total additional heat to be added to the base value of specific heat of dry (dehydrated) gypsum can be calculated as $2260 \times e \times f$, where 'f' is to take into consideration the effects of water movement and breakage of chemical bonds to release water. The value of 'f' is approximately 1.8 according to Ang and Wang (2009). The average increase in specific heat of gypsum can then calculated as $2260 \times e \times f / \Delta T$ where ΔT is the temperature range during which dehydration happens, which can be taken as 100°C starting at 100°C and ending at 200°C. The distribution of this additional specific heat has little influence on steel temperatures of interest, although a triangular distribution may be used for convenience. Therefore, the additional specific heat of gypsum due to dehydration is 0 at 100°C, linearly increasing to $2 \times 2260 \times e \times f / \Delta T$, where $\Delta T = 100$°C at 150°C and then linearly decreasing to 0 at 200°C.

5.3.2.3 Thermal conductivity

Water has much higher thermal conductivity than air. Therefore, after dehydration, the thermal conductivity of gypsum is lower than before dehydration. However, at high temperatures after dehydration, the thermal conductivity of gypsum increases with increasing temperature. There are two schools of thought on why the thermal conductivity of gypsum increases with increasing temperature. Rahmanian and Wang (2012) attributed this to void radiation after dehydration, while Ghazi Wakili et al. (2015) thought that this was induced by a type of sintering process (softening of the contacts between individual crystal needles), which increases the total contact surface and hence the conduction of heat through the bulk material.

Whatever the cause, they suggested similar thermal conductivity–temperature relations for gypsum. The relationship is as follows:

- Ambient temperature value maintained until 100°C,
- Linearly decreasing to a value of 0.1 W/(m·C) until 200°C,
- Linearly increasing to a value to that of ambient temperature until 1000°C.

5.4 ADDITIONAL ISSUES

When loaded thin-walled steel panels with gypsum plasterboard facing and interior insulation are exposed to fire attack, the gypsum plasterboard may crack and fall-off and the interior insulation may shrink or detach from each other to form holes, as shown in Figure 5.1. This will have effects on the temperature developments in the steel structure. At present, this phenomenon has

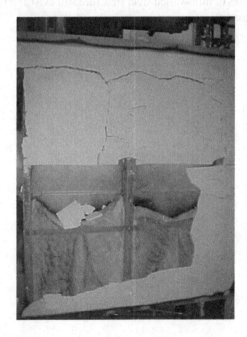

FIGURE 5.1 Behaviour of gypsum and interior insulation after a fire test. (Reprinted from *Fire Saf. J.*, 40, Feng, M. and Wang, Y.C., An experimental study of loaded full-scale cold-formed thin-walled steel structural panels under fire conditions, 43–63, Copyright 2005, with permission from Elsevier.)

been observed by many researchers, but it is still not possible to predict when this would happen. This usually happens later in fire tests and it is possible that this occurrence is induced by the failure of the steel structure. Should this be the case, then fire resistance of the construction is not affected.

5.5 SUMMARY

The accuracy of any analysis depends on the accuracy of input of the material properties. For the evaluation of fire resistance of structures, accurate data of both thermal and mechanical properties of the materials are necessary. For thin-walled steel structures, this chapter has recommended temperature-dependent thermal and mechanical property values for some protective materials and steel, respectively. However, significant gaps in knowledge on the material properties of thin-walled steel structure still exist and further research studies are necessary.

Performance-Based Design Methods of Thin-Walled Steel Members at Elevated Temperatures

6

Thin-walled steel members usually form part of a panel system, either as part of walls or floors. When designing for fire resistance, these panels are assumed to be exposed to fire from one side. Therefore, temperature distributions in thin-walled steel members are non-uniform, with steep gradients. Thin-walled steel members are also prone to different modes of buckling, including local buckling, distortional buckling and global buckling. Therefore, the behaviour of thin-walled steel members in fire is very complicated. It is not surprising that so far there is no widely accepted design calculation method for thin-walled steel members in fire.

This chapter will present a number of different methods that are either in design codes and standards or were a result of extensive research, and explain their scopes of application. The authors hope that they become the basis of further assessment by interested readers who may wish to develop a design method for their applications.

6.1 THIN-WALLED STEEL MEMBERS WITH UNIFORM TEMPERATURE DISTRIBUTION

As mentioned above, thin-walled steel members experience non-uniform temperature distributions in their sections in the majority of applications. Therefore, this chapter will primarily focus on this case.

In the case that the temperature distribution in the thin-walled steel section is uniform, the existing methods of design for thin-walled steel members at ambient temperature, such as the effective width method in Eurocode (EN 1993-1-3, CEN 2006a) or the direct strength method in North American Specification AISI S100-16 (AISI 2016) and Australia/New Zealand Standard AS/NZS 4600 (SA 2018), may be used, provided the ambient temperature mechanical properties of steel are replaced by those at elevated temperatures.

Alternatively for simple estimations, the limiting temperature method may be used. The results of a numerical investigation by Maia et al. (2016) suggest that limiting temperatures of 600°C, 550°C, 500°C and 400°C may be used for load ratios of 0.3, 0.4, 0.5 and 0.6, respectively, the load ratio being defined as the applied load in fire to the load carrying capacity of the member at ambient temperature. This limiting temperature – time relationship is broadly similar to the temperature-0.2% proof stress (yield stress) reduction factor of cold-formed steel, given in Chapter 5. These limiting temperatures are higher than the default value of 350°C, which is the currently recommended value in Eurocode 3 Part 1.2 (CEN 2005) and in Australian standard AS/NZS 4600:2018 (SA 2018).

6.2 SIMPLIFIED METHODS FOR THIN-WALLED MEMBERS WITH NON-UNIFORM TEMPERATURE DISTRIBUTION

6.2.1 Limiting Temperature Method

The limiting temperature method has already been mentioned and existing research studies indicate that this method is reasonable for uniformly heated thin-walled steel members. However, for thin-walled steel members with non-uniform temperature distribution in the cross-section, Feng et al. (2003d) found that the limiting temperature method is no longer applicable, due mainly to the influence of thermal bowing. In EN 1993-1-2 (CEN 2005) and Australian/New Zealand standard AS/NZS 4600:2018 (SA 2018), the recommended limiting temperature is 350°C. The results of Feng et al. (2003d) confirm that this value is safe provided this value is applied to the maximum temperature of the steel section. The average temperatures in thin-walled steel members at failure may be lower than 350°C.

However, applying a limiting temperature value of 350°C to the maximum temperature of the steel cross-section is unduly conservative in many cases.

6.2.2 Extension of Fire Test Results

The previous discussions have focused on resistance of thin-walled steel members at elevated temperatures, which is only part of the process of assessing their fire resistance. The other aspect is thermal response. Therefore, a complete design calculation method would need to be able to accurately predict the thermal responses of thin-walled steel members. Such a method has been described in Chapter 4, but the calculation results using such methods may not be sufficiently accurate because thin-walled steel panel structures typically contain many components and there are uncertainties in their properties, for example thermal properties and falling off of gypsum boards and thermal properties of the interior or external insulation. Therefore, fire resistance testing still represents the dominant way of demonstrating acceptance of thin-walled steel member panel systems in practice.

However, fire resistance tests are time-consuming and expensive to conduct and the dimensions of structures are limited by the dimensions of the fire test furnaces. Therefore, only a limited number of fire resistance tests are conducted for each system and the construction used in practice may be different from those tested.

To allow different constructions of the same system to be used, an approach is to extend the scope of application of the fire test results. Provided the extension is over a limited range, this may be possible and one simple method has recently been developed by the UK's Steel Construction Institute (SCI 2019).

6.2.2.1 Members in walls

For thin-walled steel members in panels under axial load, the starting point of the extension method is:

$$N_{Rd,f} = 0.6 N_{b,Rd} SRF \left(T_{ref} \right) \tag{6.1}$$

where:

$N_{Rd,f}$ is the load carrying capacity of the thin-walled steel member at the fire limit state

$N_{b,Rd}$ is the load carrying capacity of the member in normal conditions

T_{ref} is the reference temperature of the thin-walled steel section, which is taken as the mean temperature of the section. This may be taken as the average of the flange temperatures.

$SRF(T_{ref})$ is the strength reduction factor for cold-formed steel at T_{ref} according to BS EN 1993-1-2 (2005). Because the extension method is based on having availability of a valid fire test, the steel temperatures are from the fire test. In fire resistance tests on thin-walled steel panels, the thinnest steel section is usually used. Therefore, the test temperatures can be applied to thicker steel sections, which will have lower temperatures.

The multiplication factor of '0.6' is to allow for the effects of thermal bowing due to non-uniform temperature distribution in the steel cross-section.

This method allows the fire test result to be used for steel members that are slightly longer (+10% in height), thicker steel members that resist higher loads at the same fire resistance time, or to allow steel members to resist higher loads at lower fire resistance times for which temperature data would be available. However, this method is not applicable to steel members with reduced load level to achieve longer fire resistance times for which experimental data of temperatures would not be available. Due to gross approximations, this method should only be extended (extrapolated) over a small range, e.g. about 10%.

6.2.2.2 Members in floors

Steel members in floor panels will be under bending so the effects of thermal bowing are negligible. Also because the steel members will be restrained by panel boards at regular intervals, global buckling behaviour of the steel members is controlled by spacing of the restraints. Therefore, it may be assumed that the steel member resistance to bending is independent of the span. Consequently, the same fire resistance test result for one span can be used for others provided the maximum bending moments in the members are the same. As with the extension method for vertical members, the extension method can be used for thicker steel sections and for fire resistance times no longer than the test fire resistance time.

6.3 THIN-WALLED STEEL COLUMNS WITH NON-UNIFORM TEMPERATURE DISTRIBUTION IN THE CROSS-SECTION

6.3.1 Effective Width Method

In Europe, the effective width method is used to calculate the resistances of thin-walled steel members at ambient temperature. Based on the results of a relatively limited set of parametric studies, Feng et al. (2003d) evaluated modification of the effective width method for elevated temperature applications. Due to non-uniform temperature distribution in the cross-section, a number of assumptions had to be made, including the following:

- The shear centre and warping constant of the cross-section are the same as at ambient temperature. The weighted average stiffness value of the cross-section is used.
- The effective width of the web is calculated using the weighted average steel stiffness value. This affects the location of the centroid and stress distribution of the cross-section.
- The cross-section resistance of the effective section is either governed by first yield, i.e. whenever the compression or tensile stress reaches the corresponding elevated temperature yield stress the earliest, or partial plasticity. Partial plasticity happens when the tensile side reaches yield and extends the region of plasticity until the maximum compressive stress reaches yield.
- Due to the change of centre of resistance of the cross-section with non-uniform temperature distribution and thermal bowing, bending moments are present in axially loaded members. Therefore, the thin-walled member should be evaluated under combined compression and bending. The interaction equations in EN 1993-1-3 (CEN 2006a) may be used.

The results of Feng et al. (2003d) indicate that it is possible to use the effective width method to calculate the resistances of thin-walled steel members in wall construction with a reasonable level of accuracy. Using the first yield condition to calculate the cross-section resistance would give lower resistances than using the partial plasticity condition, but overall differences are small.

This assessment of the effective method was performed for a limited number of cases and there was no detailed consideration of different failure modes (local, distortional, global buckling). For thin-walled sections with internal web stiffeners, the approach of using the weighted average stiffness value of the whole web will not be appropriate, instead, the weighted average values between adjacent stiffeners should be used. The calculation procedure would be rather tedious to implement by hand, so a bespoke computer program should be written.

6.3.2 Direct Strength Method

The direct strength method (DSM) for the design of thin-walled structures is now widely adopted in design standards of many countries, including North American (AISI S100-16) and Australia/New Zealand (AS/NZS 4600:2018). It uses buckling equations to combine the cross-sectional resistances of thin-walled member with elastic critical loads of the member to calculate the design resistance, similar to design calculations for steel members with thicker sections. The elastic critical loads can be easily calculated using a computer program, one of which is CUFSM which is freely available. When calculating the plastic cross-section resistance (squash load), the gross cross-section dimensions are used.

Compared with the effective width method, DSM has advantages including being able to deal with interactions of different buckling modes and a streamlined calculation procedure. For thin-walled members under compression in wall panel applications, Shahbazian and Wang (2014a) carried out an extensive assessment of DSM. The results of their investigations confirm that DSM can be applied for thin-walled steel members with both uniform and non-uniform temperature distributions in the cross-sections for all three buckling modes. Due to temperature effects, modifications to the ambient temperature DSM equations are necessary. This method will be summarised below.

For thin-walled members with non-uniform temperature distribution under compression, additional bending moments will be generated due to thermal bowing and shift of centre of resistance of the cross-section. To make use of ambient temperature equations for thin-walled steel members under compression, Shahbazian and Wang (2014a) treated a thin-walled steel member under combined compression and bending as an equivalent column under compression only. The cross-section resistance of the equivalent column is the same as the axial force in the cross-section plastic compression-bending resistance interaction curve under the same bending moment as caused by thermal bowing (at member centre) or shift of centre of resistance (at member ends). The bending resistance is taken about the centre of resistance with non-uniform temperature distribution. At the member ends, thermal bowing is zero so there is only shift of centre of resistance. At column mid-height, both thermal bowing and shift of centre of resistance exist and they act in opposite direction. The thermal bowing is calculated using Eq. 6.2. This is illustrated in Figure 6.1 (Shahbazian and Wang 2014a).

The elastic critical loads of a thin-walled member with non-uniform temperature distribution can be calculated using CUFSM, with the different parts of the cross-section having different Young's moduli at different elevated temperatures. The shift of centre of resistance is automatically included by inputting different Young's moduli of steel. To include the effect of thermal bowing,

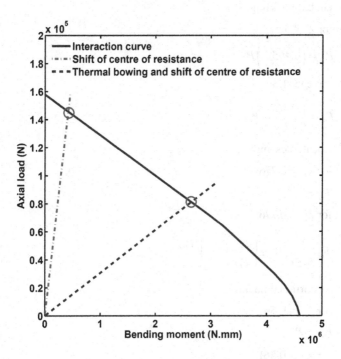

FIGURE 6.1 Determination of effective squash load of effective column. (From Shahbazian, A. and Wang, Y.C., *Struct. Eng.*, 92, 52–62, 2014.)

an additional bending moment is applied to the member, which is equal to the applied compression force acting at an eccentricity equal to the thermal bowing of the member calculated below:

$$\delta_m = \frac{\alpha L^2 \Delta T}{8d} \tag{6.2}$$

where:
    α is the coefficient of thermal expansion of steel
    L is the column length (panel height)
    ΔT is the temperature difference between the exposed and unexposed
        sides
    d is depth of the cross-section

The DSM equations recommended by Shahbazian and Wang (2014a) are:

- Global buckling:

  for $\lambda_e \leq 1.5$:

  $$P_{ne} = \left(0.495^{\lambda_e^2}\right) P_y \tag{6.3}$$

  for $\lambda_e > 1.5$:

  $$P_{ne} = \left(\frac{0.462}{\lambda_e^2}\right) P_y \tag{6.4}$$

- Local buckling:

  for $\lambda_l \leq 0.776$:

  $$P_{nl} = P_{ne} \tag{6.5}$$

  for $\lambda_l > 0.776$:

  $$P_{nl} = \left(1 - 0.22\left(\frac{P_{crl}}{P_{ne}}\right)^{0.75}\right)\left(\frac{P_{crl}}{P_{ne}}\right)^{0.75} P_{ne} \tag{6.6}$$

- Distortional buckling:

  for $\lambda_d \leq 0.561$:

  $$P_{nd} = P_y \tag{6.7}$$

  for $\lambda_d > 0.561$:

  $$P_{nd} = 0.65\left(1 - 0.14\left(\frac{P_{crd}}{P_y}\right)^{0.7}\right)\left(\frac{P_{crd}}{P_y}\right)^{0.7} P_y \tag{6.8}$$

In the above formulas P_y is the equivalent squash load of the cross-section; P_{cre} is the critical elastic global buckling load; P_{crl} is the critical elastic local buckling load; P_{crd} is the critical elastic distortional buckling load; λ_e, λ_l and λ_d are the slenderness for global, local and distortional buckling modes, respectively; P_{ne}, P_{nl} and P_{nd} are the column axial strength for global, local and distortional buckling modes, respectively; P_n is the final design strength under axial compression.

The DSM equations above have been demonstrated to be applicable for thin-walled members in wall panels under both the standard and parametric design fire curves of EN 1991-1-2 (2005).

6.3.3 Simplified Effective Width/Direct Strength Method

The only national standard that includes a complete 'performance' based design method for thin-walled steel members is the Australian/New Zealand standard AS/NZS 4600:2018 (SA 2018). This standard recommends using either the effective width method or the direct strength method, as summarised below.

6.3.3.1 Effective width method

The effective width method is used to calculate the cross-section resistances of thin-walled steel walls. The effective section is calculated using the elevated temperature mechanical properties. The effects of non-uniform temperatures are considered when calculating 'effective' average stresses for different cross-section resistances. For calculating the cross-section squash load, the weighted average yield stress of the gross cross-section is used. The cross-section bending resistances are the elastic moment resistances. At the member ends where the hot flange is under compression due to shift of centre of resistance, the yield stress of the hot flange is used. At the member centre where the cold flange is under compression due to thermal bowing, the yield stress at mid-web temperature is used.

The effective width method is combined with second-order member analysis in which the design bending moment at the member centre is magnified using the magnification factor $\frac{1}{\left(1-\frac{N_{Ed}}{N_{cr,T}}\right)}$, in which N_{Ed} is the design axial compression load in the member and $N_{cr,T}$ the flexural axial buckling capacity. $N_{cr,T}$ is calculated using the flexural stiffness of the cross-section taking into consideration the non-uniform distribution of Young's modulus of steel in the cross-section due to non-uniform temperature distribution and shift of neutral axis caused by the variation of Young's modulus.

A linear moment-axial load interaction relationship is then used to check whether the member has adequate load carrying capacity.

6.3.3.2 Direct strength method

When using the direct strength method, the axial compression and bending resistances of the member are calculated using the direct strength method under individual actions. The direct strength equations are the same as those in AISI S100-16 (AISI 2016) at ambient temperature but the gross cross-section resistances (bending and compression) and critical buckling loads are modified from those at ambient temperature to take into consideration the effects of elevated temperatures. For calculating the critical buckling loads, the critical global flexural buckling resistance is that of the gross cross-section with non-uniform temperature distribution. The critical local buckling resistance (bending or compression) is modified by using the Young's modulus reduction factor at the mid-web temperature. Distortional buckling is not considered. The critical bending resistances are obtained by modifying the ambient temperature values of the gross cross-section using Young's modulus reduction factors at the reference temperatures. For the member ends, the reference temperature is that of the hot flange (compression flange) and for the member mid-height, the reference temperature is that of the mid-web.

A linear interaction equation between axial compression and bending is then used to check whether the member has sufficient load carrying capacity. In the interaction equation, the design bending moment is calculated in the same way as in the effective width method, by considering member second-order effects. Compared to the effective width method, the differences are that the direct strength resistances of the member with gross cross-section under compression and bending separately are used in the direct strength method, instead of the resistances of the effective cross-section.

Compared with the effective width method of Feng et al. (2003d) and the direct strength method of Shahbazian and Wang (2014a), the AS/NZS 4600:2018 (SA 2018) method appears to be simpler for implementation by hand, as this method makes use of only selected reference temperatures to modify the ambient temperature calculation equations to obtain each respective elevated temperature value.

6.4 THIN-WALLED STEEL BEAMS WITH NON-UNIFORM TEMPERATURE DISTRIBUTION IN THE CROSS-SECTION

For thin-walled steel members in floors, the steel member is restrained by floor boards so there is no global lateral torsional buckling. Also there is no axial load in the member, therefore the shift of centre of resistance and thermal bowing, which dominate the behaviour and resistance of thin-walled members in wall panels, do not affect the member resistance.

AS/NZS 4600:2018 (SA 2018) appears to be the only standard that includes a design method for thin-walled members in floors. Again, both the effective width method and the direct strength method can be used. In both methods, the steel section with non-uniform temperature distribution is treated as that with uniform temperature with the mid-web temperature being taken as the reference uniform temperature of the cross-section. When using the effective width method, the resistance of the member is that at first yield of the effective cross-section.

6.5 ILLUSTRATIVE DESIGN EXAMPLE USING DIRECT STRENGTH METHOD

Requirement: calculate the load carrying capacity of a panel at 60 min of the standard fire exposure time.

Basic input data:
Steel section dimensions: lipped channel 75 × 50 × 15 × 2.5
Interior insulation: ISOWOOL
Gypsum: One layer of 12.5 mm thick Fireline British gypsum on each side
Panel height: 3 m

The calculated temperatures of the exposed and unexposed flanges (following the procedure described in Section 4.2.4) are 510.42°C and 273.85°C, respectively.

MECHANICAL PROPERTIES OF STEEL AT AMBIENT TEMPERATURE (N/mm²)	
Elastic modulus (E)	Yield stress (f_y)
205,000	350

Using program CALESL (http://j.mp/CALESL), the centre of plastic resistance is 31.3 mm from the lower temperature flange side (37.5 mm at ambient temperature). Therefore, the shift of centre of resistance is 37.5 − 31.3 = 6.2 mm. Using Eq. 6.2 gives a thermal bowing value of 49.6 mm. Therefore, the eccentricities at the top and bottom, and at the centre of the column are 6.2 mm and $|6.2 - 49.6| = 43.4$ mm, respectively.

Using program CALESL, the column axial load-bending moment interaction curve is obtained, as shown in Figure 6.2. From the two eccentricities above, the effective squash loads are 98.29 kN (top and bottom) and 49.41 kN (middle), respectively. Therefore, the effective squash load is min(98.29, 49.41) = 49.41 kN.

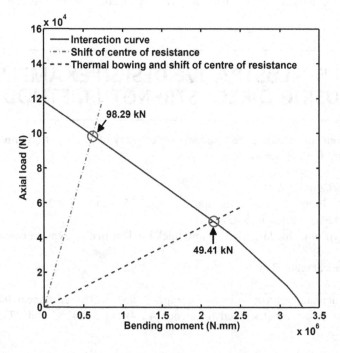

FIGURE 6.2 Determination of the effective squash load. (From Shahbazian, A. and Wang, Y.C., *Struct. Eng.*, 92, 52–62, 2014.)

Using the CUFSM (http://bit.ly/CUFSM) program, with input of the elevated temperature Young's modulus values of the steel section, the elastic critical loads for global, distortional and local buckling are:

$P_{cre} = 68.04$ kN
$P_{crl} = 388.85$ kN
$P_{crd} = 317.7$ kN

Using Eqs. 6.3–6.8:

$\lambda_c = 0.8522; P_{ne} = 29.65$ kN (Eq. 6.4)
$\lambda_l = 0.2761; P_{nl} = 29.65$ kN (Eq. 6.6)
$\lambda_d = 0.3944; P_{nd} = 49.41$ kN (Eq. 6.8)
$P_n = \min(29.65, 29.65, 49.41) = 29.65$ kN

6.6 SUMMARY

This chapter has presented an outline of a number of methods of calculating the load carrying capacity of thin-walled steel structures at elevated temperatures. Thin-walled steel structures have non-uniform temperature distributions and complex modes of failure. Therefore, there is still no universally accepted design calculation method. Nevertheless, as have been demonstrated in this chapter, it is possible to use the existing ambient temperature design methods as the basis for further development.

References

Alfawakhiri, F. (2001). Behaviour of cold-formed steel framed walls and floors in standard fire resistance tests. PhD thesis, Carleton University, Ottawa, Canada.

Alfawakhiri, F. and Sultan, M.A. (2001). Loadbearing capacity of cold-formed steel joists subjected to severe heating. *Proceedings of the 9th International Conference Proceedings, Interflam 2001*, Edinburgh, UK, Vol. 1, pp. 431–442.

Alfawakhiri, F., Sultan, M.A. and MacKinnon, D.H. (1999). Fire resistance of load bearing steel-stud walls protected with gypsum board. *Journal of Fire Technology*, Vol. 35, pp. 308–335.

American Iron and Steel Institute (AISI) (2016). S100-16: North American specification for the design of cold-formed steel structural members, 2016 edition, AISI, Washington, DC.

Anapayan, T. and Mahendran, M. (2012). Improved design rules for hollow flange sections subject to lateral distortional buckling. *Thin-Walled Structures*, Vol. 50, pp. 128–140.

Ang, C.N. and Wang, Y.C. (2009). Effect of moisture transfer on specific heat of gypsum plasterboard at high temperatures. *Construction and Building Materials*, Vol. 23, pp. 675–686.

Ariyanayagam, A. and Mahendran, M. (2018). Fire performance of load bearing LSF walls made of low strength steel studs. *Thin-Walled Structures*, Vol. 130, pp. 487–504.

Ariyanayagam, A. and Mahendran, M. (2019). Influence of cavity insulation on the fire resistance of light gauge steel framed walls. *Construction and Building Materials*, Vol. 203, pp. 687–710.

Ariyanayagam, A.D. and Mahendran, M. (2017). Fire tests of non-load bearing light gauge steel frame walls lined with calcium silicate boards and gypsum plasterboards. *Thin-Walled Structures*, Vol. 115, pp. 86–99.

Ariyanayagam, A.D., Kesawan, S. and Mahendran, M. (2016). Detrimental effects of plasterboard joints on the fire resistance of light gauge steel frame walls. *Thin-Walled Structures*, Vol. 107, pp. 597–611.

Avery, P., Mahendran, M. and Nasir, A. (2000). Flexural capacity of hollow flange beams. *Journal of Constructional Steel Research*, Vol. 53, pp. 201–223.

Babrauskas, V. (1979). COMPF2, a program for calculating post-flashover fire temperatures (Vol. 991). Department of Commerce, National Bureau of Standards.

Baleshan, B. and Mahendran, M. (2016). Experimental study of LSF floor systems under fire conditions. *Advances in Structural Engineering*. doi:10.1177/1369433216653508.

Bandula Heva, Y. and Mahendran, M. (2013). Flexural-torsional buckling tests of cold-formed steel compression members at elevated temperatures. *Steel and Composite Structures*, Vol. 14, No. 3, pp. 205–227.

Baspinar, M. and Kahraman, E. (2011). Modifications in the properties of gypsum construction element via addition of expanded macroporous silica granules. *Construction and Building Materials*, Vol. 25, pp. 3327–3333.

Baux, C., Mélinge, Y. and Lanos, C. (2008). Enhanced gypsum panels for fire protection. *Journal of Materials in Civil Engineering*, Vol. 20, pp. 71–77.

The Building Regulations 2010 (BR2010). Approved Document B: Fire safety, 2006 edition incorporating 2010 and 2013 amendments, HM Government, London, UK, 2013.

Cadorin, J.F., Pintea, D., Dotreppe, J.C. and Franssen, J.M. (2003). A tool to design steel elements submitted to compartment fires—OZone V2. Part 2: Methodology and application. *Fire Safety Journal*, Vol. 38, No. 5, pp. 429–451.

Cengel, Y. and Ghajar A.J. (2014). *Heat and Mass Transfer: Fundamentals and Applications*. McGraw-Hill Higher Education, New York.

Chen, J. and Young, B. (2006). Corner properties of cold-formed steel sections at elevated temperatures. *Thin-Walled Structures*, Vol. 44, pp. 216–223.

Chen, J. and Young, B. (2007). Experimental investigation of cold-formed steel material at elevated temperatures. *Thin-Walled Structures*, Vol. 45, pp. 96–110.

Chen, W. and Ye, J. (2014). Fire resistance prediction of load bearing cold-formed steel walls lined with gypsum composite panels. *Proceedings of 22nd International Specialty Conference on Cold-formed Steel Structures*. St. Louis, MO, pp. 541–555.

Chen, W. and Ye, J.H. (2012). Mechanical properties of G550 cold-formed steel under transient and steady state conditions. *Journal of Constructional Steel Research*, Vol. 73, pp. 1–11.

Chen, W., Ye, J., Bai, Y. and Zhao, X.L. (2012). Full-scale fire experiments on load-bearing cold-formed steel walls lined with different panels. *Journal of Constructional Steel Research*, Vol. 79, pp. 242–254.

Chen, W., Ye, J., Bai, Y. and Zhao, X.L. (2013a). Improved fire resistant performance of load bearing cold-formed steel interior and exterior wall systems. *Thin-Walled Structures*, Vol. 73, pp. 145–157.

Chen, W., Ye, J., Bai, Y. and Zhao, X.L. (2013b). Thermal and mechanical modelling of load-bearing cold-formed steel wall systems in fire. *ASCE Journal of Structural Engineering*, Vol. 140, p. A4013002.

Cooper, L.Y. and Forney, G.P. (1990). The Consolidated Compartment Fire Model (CCFM) computer code application CCFM. Vents-Part III: Catalog of algorithms and subroutines. Report NISTIR 4344, National Institute of Standards and Technology, Gaithersburg, MD.

Craveiro, H.D., Rodrigues, J.P.C., Santiago, A. and Laim, L. (2016). Review of the high temperature mechanical and thermal properties of the steels used in cold formed steel structures – The case of the S280Gd+Z steel. *Thin-Walled Structures*, Vol. 98, pp. 154–168.

Dempsey, R.I. (1990). Structural behaviour and design of hollow flange beams. *National Structural Engineering Conference*, Adelaide, Australia.

Dias, Y., Keerthan, P. and Mahendran, M. (2019a). Fire performance of steel and plasterboard sheathed non-load bearing LSF walls. *Fire Safety Journal*, Vol. 103, pp. 1–18.

Dias, Y., Mahendran, M. and Keerthan, P. (2018). Predicting the fire performance of LSF walls made of web stiffened channel sections. *Engineering Structures*, Vol. 168, pp. 320–332.

Dias, Y., Mahendran, M. and Keerthan, P. (2019b). Full-scale fire tests of steel and plasterboard sheathed web-stiffened stud walls. *Thin-Walled Structures*, Vol. 137, pp. 81–93.

European Committee for Standardization (CEN) (2002). Eurocode 1: Actions on structures – Part 1–2: General actions – Actions on structures exposed to fire. European Committee for Standardization, Brussels, Belgium.

European Committee for Standardization (CEN) (2005). EN 1993-1-2, Eurocode 3: Design of steel structures – Part 1.2: General rules – Structural fire design. European Committee for Standardization, Brussels, Belgium.

European Committee for Standardization (CEN) (2006a). EN 1993-1-3 2006, Eurocode 3: Design of steel structures – Part 1–3: General rules – Supplementary rules for cold-formed members and sheeting. European Committee for Standardization, Brussels, Belgium.

European Committee for Standardization (CEN) (2006b). EN 1993-1-5: Eurocode 3: Design of steel structures – Part 1–5: Plated structural elements. British Standards Institution, London, UK.

Feng, M. (2003). Numerical and experimental studies of cold-formed thin-walled steel studs in fire. PhD thesis, University of Manchester.

Feng, M. and Wang, Y.C. (2005). An experimental study of loaded full-scale cold-formed thin-walled steel structural panels under fire conditions. *Fire Safety Journal*, Vol. 40, pp. 43–63.

Feng, M., Wang, Y.C. and Davies, J.M. (2003a). Structural behaviour of cold-formed thin-walled short steel channel columns at elevated temperatures, part 1: Experiments. *Thin-Walled Structures*, Vol. 41, No. 6, pp. 543–570.

Feng, M., Wang, Y.C. and Davies, J.M. (2003b). Structural behaviour of cold-formed thin-walled short steel channel columns at elevated temperatures, part 2: Design calculations and numerical analysis. *Thin-Walled Structures*, Vol. 41, No. 6, pp. 571–594.

Feng, M., Wang, Y.C. and Davies, J.M. (2003c). Thermal performance of cold-formed thin-walled steel panel systems in fire. *Fire Safety Journal*, Vol. 38, No. 4, pp. 365–394.

Feng, M., Wang, Y.C. and Davies, J.M. (2003d). Axial strength of cold-formed thin-walled steel channels under non-uniform temperatures in fire. *Fire Safety Journal*, Vol. 38, pp. 679–707.

Franssen, J.M. and Gernay, T. (2017). Modeling structures in fire with SAFIR®: Theoretical background and capabilities. *Journal of Structural Fire Engineering*, Vol. 8, No. 3, pp. 300–323.

Gerlich, J.T., Collier, P.C.R. and Buchanan, A.H. (1996). Design of light steel-framed walls for fire resistance. *Fire and Materials*, Vol. 20, No. 2, pp. 79–96.

Ghazi Wakili, K., Koebel, M., Glaettli, T. and Hofer, M. (2015). Thermal conductivity of gypsum boards beyond dehydration temperature. *Fire and Materials*, Vol. 39, pp. 85–94.

Gunalan, S., Bandula Heva, Y. and Mahendran, M. (2014). Flexural-torsional buckling behaviour and design of cold-formed steel compression members at elevated temperatures. *Engineering Structures*, Vol. 79, pp. 149–168.

Gunalan, S., Bandula Heva, Y. and Mahendran, M. (2015). Local buckling studies of cold-formed steel compression members at elevated temperatures. *Journal of Constructional Steel Research*, Vol. 108, pp. 31–45.

Gunalan, S., Kolarkar, P.N. and Mahendran, M. (2013). Experimental study of load bearing cold-formed steel wall systems under fire conditions. *Thin-Walled Structures*, Vol. 65, pp. 72–92.

Ha, T.H., Cho, B.H., Kim, H. and Kim, D.J. (2016). Development of an efficient steel beam section for modular construction based on six-sigma. *Advances in Material Science and Engineering*, Vol. 2016, Article ID 9687078. doi:10.1155/2016/9687078.

Hanna, M.T., Abreu, J.C.B., Schafer, B.W. and Abu-Hamd, M. (2015). Post-fire buckling strength of CFS walls sheathed with magnesium oxide or ferrocement boards. *Proceedings of the Annual Stability Conference*, Structural Stability Research Council, Nashville, TN, March 24–27, 2015.

Iding, R., Bresler, B. and Nizamuddin, Z. (1977). FIRES-T3, A computer program for the fire response of structures-thermal. Report No. UCB FRG 77, 15.

ISO 834-1 (1999). Fire resistance tests – Elements of building construction, Part 1: General requirements. International Organization for Standardization, Geneva, Switzerland.

Jatheeshan, V. and Mahendran, M. (2016a). Experimental study of cold-formed steel floors made of hollow flange channel section joists under fire conditions. *ASCE Journal of Structural Engineering*, Vol. 142, No. 2, p. 04015134.

Jatheeshan, V. and Mahendran, M. (2016b). Fire resistance of LSF floors made of hollow flange channels. *Fire Safety Journal*, Vol. 84, pp. 8–24.

Kaitila, O. (2002). Imperfection sensitivity analysis of lipped channel columns at high temperatures. *Journal of Constructional Steel Research*, Vol. 58, No. 3, pp. 333–351.

Kankanamge, N.D. and Mahendran, M. (2011). Mechanical properties of cold-formed steels at elevated temperatures. *Thin-Walled Structures*, Vol. 49, pp. 26–44.

Keerthan, P. and Mahendran, M. (2011). New design rules for the shear strength of LiteSteel beams. *Journal of Constructional Steel Research*, Vol. 67, pp. 1050–1063.

Keerthan, P., Mahendran, M. and Frost, R.L. (2013). Fire safety of steel wall systems using enhanced plasterboards. *Proceedings of the 19th International CIB World Building Congress 2013*, Brisbane, Australia, pp. 1–12.

Kesawan, S. and Mahendran, M. (2015). Predicting the performance of LSF walls made of hollow flange sections in fire. *Thin-Walled Structures*, Vol. 98(A), pp. 111–126.

Kesawan, S. and Mahendran, M. (2016). Thermal performance of load-bearing walls made of cold-formed hollow flange channel sections in fire. *Fire and Materials*, Vol. 40, No. 5, pp. 704–730.

Kesawan, S. and Mahendran, M. (2017a). Fire performance of LSF walls made of hollow flange channel studs. *Journal of Structural Fire Engineering*, Vol. 8, No. 2, pp. 149–180.

Kesawan, S. and Mahendran, M. (2017b). A review of parameters influencing the fire performance of light gauge steel framed walls. *Fire Technology*. doi:10.1007/s10694-017-0669-8.

Kodur, V.K.R. and Sultan, M.A. (2006). Factors influencing fire resistance of load-bearing steel stud walls. *Fire Technology*, Vol. 42, pp. 5–26.

Kolarkar, P.N. and Mahendran, M. (2008). Thermal performance of plasterboard lined steel stud walls. *Proceedings of the 19th International Specialty Conference on Cold Formed Steel Structures 2008*, St. Louis, MO, pp. 517–530.

Kolarkar, P.N. and Mahendran, M. (2012). Experimental studies of non-load bearing steel wall systems under fire conditions. *Fire Safety Journal*, Vol. 53, pp. 85–104.

Lawson, R.M. and Way, A.G.J. (2016). *Value Benefits of Light Steel Construction, Technical Information Sheet P409*. Steel Construction Institute, London, UK.

Leng, J., Li, J., Guest, J.K. and Schafer, B.W. (2014). Shape optimization of cold-formed steel columns with fabrication and geometric end-use constraints. *Thin-Walled Structures*, Vol. 85, pp. 271–290.

Magnusson, S.E. and Thelandersson, S. (1970). Temperature-time curves for the complete process of fire development. A theoretical study of wood fuel fires in enclosed spaces. *Acta Polytechnica Scandinavica*, Civil Engineering and Building Construction Series, No. 65, CIB/CTF/72/46, Stockholm 1970.

Maia, E., Couto, C., Vila Real, P. and Lopes, N. (2016). Critical temperatures of class 4 cross-sections. *Journal of Constructional Steel Research*, Vol. 121, pp. 370–382.

MATLAB FILE EXCHANGE, program code, accessed on 2013, https://www.mathworks .com/matlabcentral/fileexchange/41085-calculations-of-temperature-through-the-wall-panel-assembly-exposed-to-fire-from-one-side-1.

McGrattan, K., Hostikka, S., McDermott, R., Floyd, J., Weinschenk, C. and Overholt, K. (2018). Fire dynamics simulator technical reference guide volume 1: Mathematical model. NIST Special Publication.

Outinen, J. (1999). Mechanical properties of structural steel at elevated temperatures. Licentiate thesis, Helsinki University of Technology, Helsinki, Finland.

Peacock, R.D., McGrattan, K.B., Forney, G.P. and Reneke, P.A. (2017). *CFAST–Consolidated Fire and Smoke Transport (Version 7) Volume 1: Technical Reference Guide*. National Institute of Standards and Technology, Gaithersburg, MD.

Pham, C. and Hancock, G. (2013). Experimental investigation and direct strength design of high-strength, complex C-sections in pure bending. *Journal of Structural Engineering, American Society of Civil Engineers*, Vol. 139, pp. 1842–1852.

Rahmanian, I. and Wang, Y.C. (2012). A combined experimental and numerical method for extracting temperature-dependent thermal conductivity of gypsum boards. *Construction and Building Materials*, Vol. 26, No. 1, pp. 707–722.

Ranawaka, T. and Mahendran, M. (2009a). Distortional buckling tests of cold-formed steel compression members at elevated temperatures. *Journal of Constructional Steel Research*, Vol. 65, pp. 249–259.

Ranawaka, T. and Mahendran, M. (2009b). Experimental study of the mechanical properties of light gauge cold-formed steels at elevated temperatures. *Fire Safety Journal*, Vol. 44, pp. 219–229.

Rusthi, M., Ariyanayagam, A.D., Mahendran, M. and Keerthan, P. (2017). Fire tests of magnesium oxide board lined light gauge steel frame wall systems. *Fire Safety Journal*, Vol. 90, pp. 15–27.

Sakumoto, Y., Hirakawa, T., Masuda, H. and Nakamura, K. (2003). Fire resistance of walls and floors using light-gauge steel shapes. *Journal of Structural Engineering*, Vol. 129, No. 11, pp. 1522–1530.

Shahbazian, A. and Wang, Y.C. (2012). Direct Strength Method for calculating distortional buckling capacity of cold-formed thin-walled steel columns with uniform and non-uniform elevated temperatures. *Thin-Walled Structures*, Vol. 53, pp. 188–199.

Shahbazian, A. and Wang, Y.C. (2013). A simplified approach for calculating temperatures in axially loaded cold-formed thin-walled steel studs in wall panel assemblies exposed to fire from one side. *Thin-Walled Structures*, Vol. 64, pp. 60–72.

Shahbazian, A. and Wang, Y.C. (2014a). A performance-based fire resistance design method for wall panel assemblies using thin-walled steel sections. *The Structural Engineer*, Vol. 92, pp. 52–62.

Shahbazian, A. and Wang, Y.C. (2014b). A fire resistance design method for thin-walled steel studs in wall panel constructions exposed to parametric fires. *Thin-Walled Structures*, Vol. 77, pp. 67–76.

Siahaan, R., Keerthan, P. and Mahendran, M. 2016. Finite element modelling of rivet fastened rectangular hollow flange channel beams subject to local buckling. *Engineering Structures*, Vol. 126, pp. 311–327.

Standards Australia (SA) (2005). Methods for fire tests on building materials, components and structures. Fire-resistance tests of elements of building construction. AS 1530.4, Sydney, Australia.

Standards Australia (SA) (2018). Australia/New Zealand standard AS/NZS 4600 cold-formed steel structures. Sydney, Australia.

Steel Construction Institute (2019). Fire resistance of light steel framing, SCI Publication P424, Steel Construction Institute, Ascot, UK.

Stern-Gottfried, J. and Rein, G. (2012). Travelling fires for structural design-part II: Design methodology. *Fire Safety Journal*, Vol. 54, pp. 96–112.

Sultan, A.M. (1996). A model for predicting heat transfer through non-insulated unloaded steel-stud gypsum board wall assemblies exposed to fire. *Fire Technology*, Vol. 32, No. 3, pp. 239–259.

Sultan, M.A., Seguin, Y.P. and Leroux, P. (1998). Results of fire tests on full-scale floor assemblies. Internal Report, National Research Council of Canada, Ottawa, Canada.

Takeda, H. and Mehaffey, J.R. (1999). WALL2D: A model for predicting heat transfer through wood-stud walls exposed to fire. *Fire and Materials*, Vol. 22, No. 4, pp. 133–140.

Wang, Y., Burgess, I., Wald, F. and Gillie, M. (2012). *Performance-Based Fire Engineering of Structures*. CRC Press, London, UK.

Wang, Y., Ying-Ji, C. and Lin, C. (2015). The performance of calcium silicate board partition fireproof drywall assembly with junction box under fire. *Journal of Advances in Materials Science and Engineering*, Vol. 125, pp. 1–12.

Ye, J.H. and Chen, W. (2013). Elevated temperature material degradation of cold-formed steels under steady- and transient-state conditions. *ASCE Journal of Materials in Civil Engineering*, Vol. 25, No. 8, pp. 947–957.

Zhao, B., Kruppa, J., Renaud, C., O'Connor, M., Mecozzi, E., Apiazu, W., Demarco, T. et al. (2005). Calculation rules of lightweight steel sections in fire situations. Technical Steel Research, European Union.

Index